Topics in Recreational Mathematics 1/2017

Editor-in-chief

Charles Ashbacher
5530 Kacena Ave
Marion, IA 52302 USA

cashbacher@yahoo.com

Artwork

Caytie Ribble

Contributor

Rachel Pollari

ISBN-13: 978-1981159765

Copyright 2017

Table of Contents

Editor's Note	4
In Memory of Solomon W. Golomb *Charles Ashbacher*	5
In Memory of Raymond Smullyan *Charles Ashbacher*	7
In Memory of Clayton W. Dodge *Charles Ashbacher*	8
Mathematical Cartoons *Caytie Ribble*	9
"Honest" Numbers in the Languages of the Native Americans of North America *Charles Ashbacher*	12
The 2012-13 NHL Lockout: Did Hockey Fans Strike Back? *Anna M. Van Kula and Paul M. Sommers*	27
Mathematical Venery Revisited *Charles Ashbacher*	36
Alphametics *Charles Ashbacher*	38
The NFL Draft: Two Models For Draft Evaluation *Andrew W. Jung and Paul M. Sommers*	42
Wordplay and Humor *Rachel Pollari and Charles Ashbacher*	56
Trimorphic Numbers Revisited *Charles Ashbacher*	57
Which Gender is Happier in the United States? *Charles Ashbacher*	59

Fun With Words *Charles Ashbacher*	62
NON-ZERO FACTORS OF 10^n *Rudolph Ondrejka* Extension to 12^n *Charles Ashbacher*	63
Abstracts of the papers in "Journal of Recreational Mathematics" Volume 1, Numbers 2, 3 and 4, 1968 *Charles Ashbacher*	65
Curiosa on 2016 and 2017 *Charles Ashbacher*	72
Integer Compositions Applied To The Probability Analysis Of Blackjack And The Infinite Deck Assumption *Jonathan Marino and David G. Taylor*	74
Book Reviews *Charles Ashbacher*	92
95 Years of Artistic Surprises with Combinatorial Color-Marked Edge-Matching Tiles featuring their latest embodiment, MiniMatch-I *Kate Jones*	100
Neutrosphic Sets and Logic Florentin Smarandache	104
Books in Recreational Mathematics by Charles Ashbacher and Associates	106

Editor's Note

This book has been long delayed due to the serious and eventually terminal illness of my father. He passed on February 19, 2017 after a lengthy illness, spending his last days in the peaceful setting of a hospice.

This book opens with three "In Memory of . . . " pieces, three very prominent mathematicians that I interacted with have passed in the last three months. They are Solomon W. Golomb, Raymond Smullyan and Clayton W. Dodge. All three made major contributions to the field of mathematics in very different ways.

The reader will find a wide variety of material in this issue, my leading material is an examination of the languages of Native Americans in North and Central America. My work is a search for "honest" numbers in those languages, doing what little I can to publicize and preserve these important segments of the human heritage.

There are two papers from Paul Sommers where he analyzes aspects of the professional sports of hockey and football. His economic perspectives on fan reaction to a work stoppage and ways to evaluate players for the draft are both entertaining and illuminating.

Additional work in my ongoing projects to keep the flame of **Journal of Recreational Mathematics** alive in the form of the abstracts to three issues of volume 1 of **JRM** are included. As I have stated before, it is my goal to complete a collection of abstracts and a full index to all 38 volumes of **JRM**.

A detailed explanation of the probabilities in the game of blackjack is the most complex paper in this collection. Games, even those where money changes hands, have always been a key component of recreational mathematics and this paper is a significant advancement in the understanding of a mainstay of casino gambling.

Short papers that are extensions of work that previously appeared in JRM, book reviews and some alphametics round out the content of this work. I hope you enjoy it and as always, feedback and comments are appreciated.

Charles Ashbacher

cashbacher@yahoo.com

In Memory of Solomon W. Golomb

Charles Ashbacher

It is with the deepest regret that I announce the passing of Solomon (Sol) W. Golomb. Few mathematicians have the intelligence and foresight to create a new area of mathematics and Sol was one of them. His book "Polyominoes," first published in 1965, codified an entire field of recreational mathematics and Sol invented the term "polyomino" in 1953. While a few problems involving polyominoes had appeared earlier, he was the one that made it a discipline.

A polyomino is a plane geometric figure formed by putting one or more square figures of the same size together by the edges. Polyominoes of any number of squares can be created, although the smaller structures are the most easy to work with. The number of polyominoes that can be created using n squares grows rapidly. There are 12 that can be made with 5 squares, 63,600 with 12 squares and approximately 7×10^{31} with 56 squares. Therefore, the number of problems is effectively infinite. Sol's series of papers that appeared in "Recreational Mathematics Magazine" in 1961-1963 are classic papers that introduced and expanded the field.

The most famous polyominoes are the pentominoes, the twelve figures that can be made with five squares. This term was also invented by Sol and many pentomino problems appeared in "Journal of Recreational Mathematics" over the years it was published. I got my start in writing extensive programs to solve recreational mathematics problems by creating programs to solve problems involving constructions using the two and three-dimensional pentominoes.

One of the beauties of the polyominoes is that no higher mathematical ability is needed to work with them. Basic problems can be given to elementary children and they can be created for any level of difficulty and the polyominoes are inexpensive to produce. I once reviewed a kit of pentomino problems designed for elementary school children to aid them in developing spatial orientation skills.

Sol is now gone but he left a legacy of mathematics that can be used to educate and entertain. I encourage all people with an interest in math to acquire and read one of the editions of "Polyominoes." It is an addictive area of mathematics that can be read and understood by everyone.

In Memory of Raymond Smullyan

Charles Ashbacher

In his long and very productive life Raymond Smullyan created some of the best logic books ever written. Even though his puzzles were often complex, they were still generally within the ability of the general reader. His basic strategy was opening with some primer puzzles that introduced the concepts and then he engaged in a series of stepwise more complex problems until he had your mind doing the mental equivalent of flipping your lips with your finger.

Smullyan died on February 6, 2017 at the age of 97. The great popularizer of mathematics Martin Gardner once described Smullyan's book "What is the Name of This Book?" in "Scientific American" as:

"The most original, most profound, and most humorous collection of recreational logic and math problems ever written."

I had the privilege of reviewing several of Smullyan's books and engaged in a small amount of personal communication. His work is some of the most challenging that you will ever read and while his books may age, his work will never grow stale or out of style. If you get a chance, check out the video of Smullyan appearing on "The Tonight Show starring Johnny Carson."

In Memory of Clayton W. Dodge

Charles Ashbacher

Although he accomplished many other things, my experience with Clayton Dodge was largely due to his work as the problem column editor of the "Pi Mu Epsilon Journal," a position he held from 1980 to 2002. As anyone that has ever edited a problem column can attest, it is a surprisingly challenging, yet rewarding job.

Each issue contained a wide variety of problems, from a basic alphametic to complex problems in geometry, calculus and number theory. I was always very excited when the latest issue arrived, both for the vanity of seeing my name in the solvers' list and to see what new problems there were to solve.

I had the privilege of personally meeting Clayton at an MAA Mathfest meeting. I was moderating a session and he had other things to do, so both of us were time-challenged. Therefore, our interaction was brief. We communicated via email on occasion for many years, every once in awhile he would send me a problem and ask for my opinion on it. On occasion, he would issue some kind of "Can you do better?" challenge. Clayton was therefore responsible for many times when I would drop everything in order to accept the "Dodge Challenge."

Clayton contributed a great deal to the field of mathematics, problem columns are a way for people to get started on mathematical research and to experience the joy of being published. There are no monetary rewards for solving problems in a journal, but the bragging rights and the ego boost are priceless. The mathematical world is spinning a little slower with his passing on January, 22 2017.

Mathematical Cartoons

Caytie Ribble

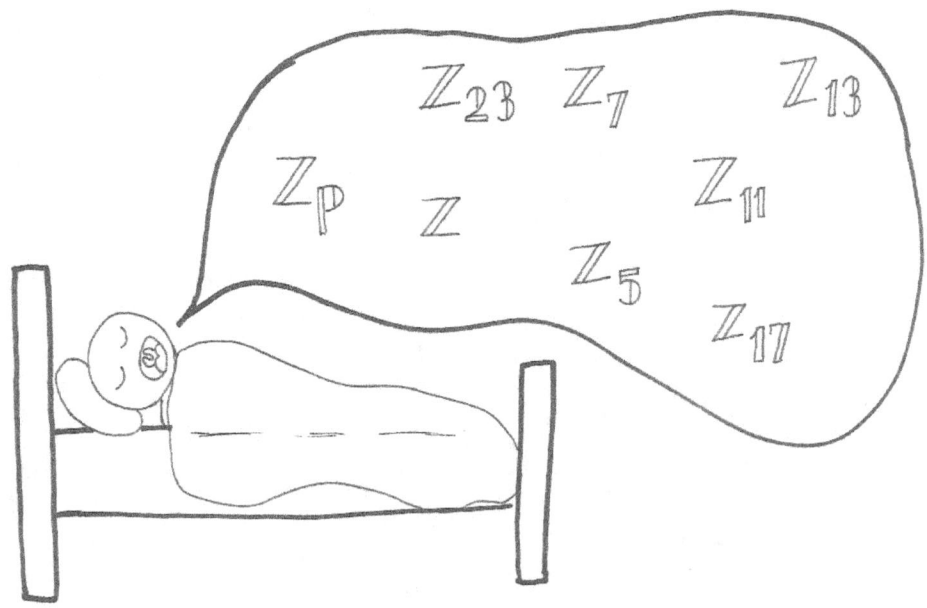

MATHEMATICAL COWS

"Honest" Numbers in the Languages of the Native Americans of North America

Charles Ashbacher

cashbacher@yahoo.com

Abstract

Like so many ideas in recreational mathematics, the concept of an "honest" number was created by Martin Gardner. A number is considered "honest" if the number of letters in the name is the value of the number. For example, "four" is the only "honest" number in English.

In this paper, the languages of Native Americans of North America are examined in a search for additional "honest" numbers.

Background

In his long-running and popular mathematical recreations column, Martin Gardner[1] defined an "honest" number as one where the number of letters in its English name matches the value. For example, "four" is an "honest" number and the only one in the English language.

In his paper "The Lucky Languages," Sidney Kravitz[2] examines 17 other western languages, looking for more "honest" numbers as well as additional characteristics.

The purpose of this paper is to search for additional "honest" numbers, this time in the languages of the Native Americans that lived in North America. The website http://www.native-languages.org is operated by Native Languages of the Americas, a non-profit organization dedicated to the preservation of Native American languages.

This site contains the numbers one through ten in the languages of the Native Americans of North America as well as Central and South America. http://www.native-languages.org/numbers.htm. Here is a list of the Native American languages of North America, as listed on the site. Some are extinct, others are in danger of extinction while some still have large numbers of native speakers.

North American Native American Languages

Abnaki-Penobscot	Adai
Achumawi	Ahtna
Alabama	Aleut
Algonquin	Alsea
Amuzgo	Apache
Arikara	Assiniboine
Atakapa	Atsugewi
Babine	Beothuk

Biloxi	Blackfoot
Caddo	Cahuilla
Cayuga	Cherokee
Chickasaw	Choctaw
Chol	Chontal Maya
Comanche	Cree
Cuicatec	Dakota
Delaware	Gabrielino
Hitchiti	Hopi
Kansa	Kaqchikel
Kaska	Koasati
Lakota	Lenape
Lushootseed	Maliseet
Matlatzinca	Maya
Miami-Illinois	Micmac
Mohegan	Montagnais
Nahuatl	Natchez
Ofo	Ojibwe
Paiute	Pame
Papago	Pawnee
Pima Bajo	Quapaw
Sauk	Shoshone
Stoney	Tanana
Timucua	Tlahuica
Tonkawa	Wichita
Yaqui	Zapotec

Native American Languages of North America Containing "Honest" Numbers

In all of the languages examined, only the numbers one through ten are considered. Therefore, it is remotely possible that there are other "honest" numbers in these languages. Furthermore, many of the letters in the languages have accent marks that may be unfamiliar to many readers. So, there will be times when it may look like the numbers are not "honest." If a language does not appear, then no "honest" numbers were discovered.

Penobscot-Abenaki is the language spoken by the Penobscot and Abenaki tribes in northern New England as well as the contiguous areas of Canada. A few of the tribal elders still speak the language. There are three "honest" numbers in **Penobscot-Abenaki** and they appear in table 1.

Table 1

Number	Word
Three	nas
Five	nôlan
Seven	dôbawôz

Adai is the language native to Northwestern Louisiana that has been extinct for centuries and the only known words are from a list of 275 words published in 1804. There are two "honest" numbers in **Adai** and they appear in table 2.

Table 2

Number	Word
Eight	pacalcon
Nine	sickinish

Alabama is a language that was native to the Southeastern United States until the members of the tribe were forcibly relocated westward. Consequently, the people that are fluent in the language are elders that live primarily in Texas. There are two "honest" numbers in the **Alabama** language and they appear in table 3.

Table 3

Number	Word
Seven	ontòklo
Nine	chákkàali

Aleut is a language native to the area of the Aleutian Islands in Alaska. It is estimated that 100-300 native speakers still exist in Alaska and Russia. There is one "honest" number in **Aleut** and it appears in table 4.

Table 4

Number	Word
six	atuung

Algonquin is a language that has several forms and was spoken nearly all the way across what is now Canada, the northern eastern seaboard of the United States and all the way to the Rocky Mountains of the plains states. Several of the dialects are extinct and others nearly so. There are two "honest" numbers in **Algonquin** and they appear in table 5.

Table 5

Number	Word
five	nànan
nine	shàngaswi

Amuzgo is a language spoken in the Costa Chica region of Mexico and has approximately 50,000 native speakers, so it is an active language. There is one "honest" number in **Amuzgo** and it appears in table 6.

Table 6

Number	Word
two	bè

Apache is a language spoken in the Southwestern United States and there are approximately 14,000 native speakers. There are two "honest" numbers in **Apache** and they appear in table 7.

Table 7

Number	Word
two	bè
four	dįį'i

Arikara is a language spoken by a tribe in the upper plains of the American Midwest. The speakers now primarily live on the Fort Berthold Reservation in North Dakota. At this time there are few fluent speakers, but it is taught in the local schools. There is one "honest" number in **Arikara** and it appears in table 8.

Table 8

Number	Word
Five	šihux

Assiniboine is a **Siouan** language of the Northern Plains of the United States. It is estimated that there are somewhere between 50 and 150 native speakers. There are three "honest" numbers in **Assiniboine** and they appear in table 9.

Table 9

Number	Word
four	doba
five	zaptą
nine	napcuwąga

Atakapa is a language that was spoken by people in Southwestern Louisiana that went extinct in the early twentieth century. There are two "honest" numbers in **Atakapa** and they appear in table 10.

Table 10

Number	Word
three	lat
six	latsīk

Babine–Witsuwit'en is a language spoken in the interior of British Columbia in Canada. It is currently considered endangered with less than 500 native speakers. There is one "honest" number in **Babine–Witsuwit'en** and it appears in table 11.

Table 11

Number	Word
five	kwile'

Beothuk is an extinct language that was spoken by people living on the island of Newfoundland in Canada. The last native speaker died in 1829 and little is known about the language. There is one "honest" number in **Beothuk** and it appears in table 12.

Table 12

Number	Word
seven	oodzook

Biloxi is a language that was spoken in the southern states of Mississippi, Louisiana and Texas. It has been extinct since the 1930's, when the last native speaker died. There are two "honest" numbers in the **Biloxi** language and they appear in table 13.

Table 13

Number	Word
four	topa'
six	akŭxpĕ'

Caddo is the only surviving language within the **Caddo** language family. It is spoken by the Caddo Nation in Oklahoma and it is endangered as there are approximately 25 native speakers remaining. There are two "honest" numbers in **Caddo** and they appear in table 14.

Table 14

Number	Word
four	hiwi'
six	dánkih

Cahuilla is a language in the **Uto-Aztecan** family spoken by people of the Cahuilla nation of Southern California. The language is listed as severely endangered, with less than fifty native speakers. There are two "honest" numbers in **Cahuilla** and they appear in table 15.

Table 15

Number	Word
three	páh
nine	qunwíchiw

Cayuga is a language in the **Iroquois** family spoken in New England and Ontario, Canada and there are approximately 200 fluent speakers remaining. There is one "honest" number in **Cayuga** and it appears in table 16.

Table 16

Number	Word
five	hwihs

Cherokee is considered to be one of the healthiest of the Native American languages, with more literature published in Cherokee than in any other Native American language. The Cherokee were originally on the East coast of the United States before most were relocated to Oklahoma. There are three "honest" numbers in **Cherokee** and they appear in table 17.

Table 17

Number	Word
four	nvgi
five	hisgi
six	sudali

Chickasaw is a language of the **Muskogean** language family that originated in the state of Mississippi. The members of the tribe were relocated to Oklahoma in the great relocation of Native Americans in the Eastern United States. The language is considered in danger of extinction with less than 100 fluent speakers remaining. There is one "honest" number in **Chickasaw** and it appears in table 18.

Table 18

Number	Word
seven	ontoklo

Choctaw is a language in the **Muskogean** family. Originally in Mississippi and other areas of the Southern United States, most were relocated to Oklahoma in the great relocation of Native Americans in the Eastern United States. There are approximately 10,000 fluent speakers, but the numbers are declining. There is one "honest" number in **Choctaw** and it appears in table 19.

Table 19

Number	Word
seven	ontuklo

Ch'ol Maya is a language in the Mayan family that is spoken by the people in the Mexican state of Chiapas. It is a dynamic language with over 200,000 native speakers. There are two "honest" numbers in **Ch'ol Maya** and they appear in table 20.

Table 20

Number	Word
seven	wukp'ejl
nine	bolomp'ejl

Comanche is a language spoken by the tribe of the same name that inhabited eastern New Mexico, southeastern Colorado, southwestern Kansas, western Oklahoma, and northwest Texas. Members of the tribe served as code talkers for the U. S. military in World War II. The language is severely endangered, with very few native speakers. There is one "honest" number in **Comanche** and it appears in table 21.

Table 21

Number	Word
nine	wumhinatu

Cree is a language in the **Algonquian** family with the largest number of native speakers of all Native American languages in Canada. The speakers are scattered across all of Canada from the Northwest Territories to Labrador. There are two "honest" numbers in **Cree** and they appear in table 22.

Table 22

Number	Word
four	newo
eight	ayinânew

Cuicatec is a language spoken in the Mexican state of Oaxaca and there are over 20,000 speakers of the language. There is one "honest" number in **Cuicatec** and it appears in table 23.

Table 23

Number	Word
three	inu

Dakota is a language spoken by the Sioux tribes of the upper midwest of the United States. There are approximately 20,000 native speakers and there are three "honest" numbers in **Dakota**. They appear in table 24.

Table 24

Number	Word
four	topa
seven	sakowin
eight	sahdogan

Hitchiti is a language that was spoken in an area of the state of Georgia on the banks of the Chattahoochee River. There are still a few native speakers and the language is in danger of extinction. There are two "honest" numbers in **Hitchiti** and they appear in table 25.

Table 25

Number	Word
six	ipa·kan
nine	u·stapakan

Kansa is a language spoken by the Kaw people of Oklahoma where the last native speaker died in 1983, although there is a program to develop second language speakers. There is one "honest" number in **Kansa** and it appears in table 26.

Table 26

Number	Word
four	tóba

Kaqchikel Maya is a language spoken by the Kaqchikel people in central Guatemala. There are hundreds of thousands of native speakers and there are two "honest" numbers in **Kaqchikel Maya**. They appear in table 27.

Table 27

Number	Word
three	oxi'
four	kaji'

Kaska is a language spoken by the Kaska people in the Yukon Territory and British Columbia in Canada. Although the language is taught in schools, the number of speakers is in the low hundreds and it is considered endangered. There is one "honest" number in **Kaska** and it appears in table 28.

Table 28

Number	Word
five	klola

Lakota is a language spoken by the Lakota people of the Sioux tribes in the northern plains of the United States. There are thousands of native speakers and it is taught in many schools. There are three "honest" numbers in **Lakota** and they appear in table 29.

Table 29

Number	Word
four	tópa
seven	sakowin
eight	saglogan

Delaware/Lenape is a term used to refer to two related languages (**Munsee** and **Unami**) spoken by the people of the east coast of the United States in the Delaware, New Jersey and New York area. **Uname** is extinct and there are less than ten native speakers of **Munsee**. There are two "honest" numbers in **Delaware/Lenape** and they appear in table 30.

Table 30

Number	Word
four	newa
five	nalan

Lushootseed is a language spoken by people in the Puget Sound area of Washington state in the United States. Although there are few native speakers, there is an effort to revive the language. There are three "honest" numbers in **Lushootseed** and they appear in table 31.

Table 31

Number	Word
three	łix^w
four	buus
five	cəlac

Malecite–Passamaquoddy, also known as **Maliseet–Passamaquoddy** is a language spoken by the people on both sides of the border between Maine in the United States and New Brunswick, Canada. It is endangered with approximately 500 native speakers. There is one "honest" number in **Malecite–Passamaquoddy** and it appears in table 32.

Table 32

Number	Word
nine	esqonatek

Matlatzinca is a term used to refer to two languages, **Ocuiltec** or **Tlahuica**, and another called **Matlatzinca**. They are mutually unintelligible and are spoken in Central Mexico. Although the number of speakers is well over a thousand, it is considered highly endangered. There is one "honest" number in **Matlatzinca** and it appears in table 33.

Table 33

Number	Word
six	ratowi

Mayan is a term used to refer to a family of languages spoken by approximately 6 million people in Central America. It is one of the most studied languages of the indigenous people of North America. There is one "honest" number in **Mayan** and it appears in table 34.

Table 34

Number	Word
three	óox

Mi'kmaq is a language of northern New England and the adjacent areas of Canada. With over 8,000 speakers, the language is not endangered. There is one "honest" number in **Mi'kmaq** and it appears in table 35.

Table 35

Number	Word
six	asikom

Mohegan/Pequot is an extinct **Algonquian** language spoken in New England as well as eastern Long Island. The last native speaker died in 1908. There are three "honest" numbers in **Mohegan/Pequot** and they appear in table 36.

Table 36

Number	Word
five	yupáw
six	qutôsk
nine	pásukokun

Nahuatl is also known as **Aztec** and is spoken by 1.5 million people in Central Mexico. There are three "honest" numbers in **Nahuatl** and they appear in table 37.

Table 37

Number	Word
three	yei
seven	chicome
ten	mahtlactli

Papago-Pima or **O'odham** is a **Uto-Aztecan** language of southern Arizona in the United States and northern Sonora in Mexico. With approximately 15,000 native speakers scattered between the two countries, the language is in no danger of becoming extinct. There is one "honest" number in **Papago-Lima** and it appears in table 38

Table 38

Number	Word
four	gi'ik

Pima Bajo is language in the **Uto-Aztecan** family with approximately 500 native speakers living in northern Mexico. There is one "honest" number in **Pima Bajo** and it appears in table 39.

Table 39

Number	Word
ten	viistamaama

Quapaw is also known as **Arkansas**, is a **Siouan** language of the Quapaw people originally from Arkansas. The native speakers currently reside in Oklahoma and the language is well-documented, but does not conform well to the phonetics of the English language. There are four "honest" numbers in **Quapaw** and they appear in table 40.

Table 40

Number	Word
four	to^{n}wa
five	sátta^n
seven	ppéno^{n}ba
eight	ppedábni^n

Stoney is also called **Nakota** and is a language in the **Siouan** family. It is spoken on reserves in Alberta, Canada and there are several thousand native speakers. There are five "honest" numbers in **Stoney** and they appear in table 41.

Table 41

Number	Word
four	ktusa
five	zaptâ
six	shakpe
seven	shagowî
eight	shaxnorâ

Tanana is a language split into **Upper Tanana** and **Lower Tanana** and is spoken in the interior of Alaska with some additional speakers in the Yukon Territory of Canada. The language is endangered with less than 100 speakers of each branch. There is one "honest" number in **Tanana** and it appears in table 42.

Table 42

Number	Word
ten	xundelaayi

Timucua was a language spoken in Central Florida and Southern Georgia that went extinct in the second half of the eighteenth century. It was the first language of Native Americans to have a writing system developed. There are four "honest" numbers in **Timucua** and they appear in table 43.

Table 43

Number	Word
five	marua
six	mareca
seven	piqicha
eight	piqinahu

Wichita is a language that was spoken in Oklahoma and recently went functionally extinct, with the last native speaker dying in 2016. However, there are efforts to revitalize the language. There is one "honest" number in **Wichita** and it appears in table 44.

Table 44

Number	Word
three	taw

Yaqui, also known as **Yoeme** is a language of the **Uto-Aztecan** family that is spoken by people in the Mexican state of Sonora as well as Arizona in the United States. With approximately 18,000 native speakers, the language is in no danger of extinction. There are three "honest" numbers in **Yaqui** and they appear in table 45.

Table 45

Number	Word
five	mamni
six	vusani
eight	wohnaiki

Zapotec is term used to refer to a family of languages spoken by people that live in the southwestern-central highlands of Mexico. With nearly a half-million native speakers, it is one of the most dynamic Native American languages. There are three "honest" numbers in **Zapotec** and they appear in table 46.

Table 46

Number	Word
four	tapa
five	gaayu'
six	xhoopa'

Summary

Of the 68 languages in the list of Native American languages on the native-languages site, 45 or roughly two-thirds contain an "honest" number in the numbers from 1 through 10. Stoney has the most with five of the numbers being "honest" while **Quapaw** and **Timucua** each have four. Ten of the languages have three "honest" numbers, thirteen have two "honest" numbers and twenty have one.

In the languages examined here, four and five are the "most honest" numbers with 16 appearances each. The number of times each of the numbers one through ten are honest numbers is summarized in table 47.

Table 47

Number	Number of times it is "honest"
1	0
2	2
3	9
4	16
5	16
6	13
7	12
8	8
9	10
10	2

References

1. Martin Gardner, "Mathematical Games," **Scientific American**, 214(1), January 1966, page 112.

2. Sidney Kravitz, "The Lucky Languages," **Journal of Recreational Mathematics**, Volume 7, page 225.

THE 2012-13 NHL LOCKOUT: DID HOCKEY FANS STRIKE BACK?

Anna M. Van Kula
Paul M. Sommers
Middlebury College
psommers@middlebury.edu

Abstract

The 2012-13 National Hockey League (NHL) lockout (the most recent work stoppage in professional sports) resulted in the cancellation of a total of 526 regular-season games. The authors assess the impact of this lockout on home game attendance for each of the thirty NHL teams in the full season after the lockout (2013-14) and one full season before the lockout (2011-12). The game-by-game 2013-14 home attendance figures compared with those a full season before the lockout reveal that only three teams experienced significantly smaller crowds in the year after the lockout. Irritation among hockey fans was not noteworthy following this particular protracted work stoppage.

The National Labor Relations Act of 1935 allows players to strike or withhold their services from owners. Owners can confront players with a lockout, refusing to allow them to play. Work stoppages — strikes (initiated by players) or lockouts (imposed by owners) — rarely happen in sports. But, when they do, neither heavily discounted tickets nor hot races mollify disgruntled fans. In 1994, a long-simmering labor dispute abruptly ended the baseball season and forced cancellation of the World Series for the first time in the history of the game. Nally and Sommers [1] show that the baseball strike in 1994-95 (the longest work stoppage in professional sports history) clearly incurred the wrath of fans. The game-by-game 1995 home attendance figures compared with those a season earlier reveal average differences that were discernibly lower for twenty-three of the total twenty-eight teams. The 2012-13 National Hockey League (NHL) lockout (the most recent work stoppage in professional sports) resulted in the cancellation of a total of 526 regular-season games. (By comparison, the 2004-05 NHL lockout led to the cancellation of the entire season.) The purpose of this brief note is to show how the attendance figures for each of the thirty NHL teams in 2013-14 changed from the lockout-shortened 2012-13 season and from the last full season (2011-12) before the lockout. All home game attendance figures are from the NHL's official website (www.NHL.com).

NHL teams played 24 home games (as opposed to the normal 41) in the lockout-shortened 2012-13 season. Table 1 shows average home attendance in 2012-13 and again in 2013-14 for each NHL team and the percent change in the two averages. For twelve of the thirty teams, the average home game attendance dropped. Among the hardest hit markets were Florida (-14.52 percent) and Carolina (-11.83 percent). Neither team qualified for the 2013-14 playoffs. The results of a series of two sample *t*-tests reported in the last column of Table 1 underscore the fans' disgust in some markets.[1] Eighteen teams, however, either experienced larger crowds or sold out both years. None of the increases was statistically significant (using $\alpha = .05$). Of the twelve teams that experienced smaller crowds in 2013-14, the drop was significant for nine of them (again, using $\alpha = .05$). And, of these nine teams, only Dallas, Minnesota, and Tampa Bay qualified for the 2013-14 playoffs.

Hockey-starved fans may have turned out in record numbers when the lockout-shortened season started on January 19, 2013. That is, given the surge in attendance after the lockout ended, any comparison between the 2012-13 and 2013-14 seasons (in Table 1) might show (not surprisingly) that attendance, on average, decreased. Table 2 shows average home attendance in 2011-12 (one full season before the lockout) and 2013-14 (one full season after the lockout). Average home attendance increased for fourteen teams from 2011-12 to 2013-14; ten more teams sold out both years. In other words, only six teams experienced smaller crowds and the drop was significant for just three of them (Florida, -12.65 percent; St. Louis, -6.97 percent; and Ottawa, -6.44 percent).

If hockey-starved fans attended games in much higher numbers during the last half of the lockout-shortened 2012-13 season, then one would expect higher average attendance figures in 2012-13 than in the season before. Table 3 shows average home attendance in 2011-12 (one full season before the lockout) and 2012-13. Average home attendance either remained the same (because they sold out all home games both seasons) or increased from 2011-12 to 2012-13 for a total of twenty-six teams. The increase was significant for ten teams; the decrease was significant for only two teams (St. Louis and Toronto).

Concluding Remarks

Work stoppages in professional sports are uncommon. When they do occur, they are resolved in time before the regular season starts. In 2004-2005, owners locked out the players, and the entire NHL season was lost. In 2012-2013, close to half of the NHL season was lost. But when play resumed, attendance surged during the second half of the 2012-13 season. The game-by-game 2013-14 home attendance figures compared with those a full season before the lockout reveal that only three teams experienced significantly smaller crowds. Fourteen teams experienced larger crowds and ten more teams sold out both seasons. One can only conclude that hockey fans are less fickle than baseball fans!

Table 1
Average Home Attendance by Team,
2012-13 and 2013-14 Seasons

Team	Average home attendance 2012-13	Average home attendance 2013-14	Percent change	p-value on difference
Anaheim	15888	16469	3.66	0.1002
Arizona	13924	13776	-1.06	0.8283
Boston	17565	17565	SOLDOUT	
Buffalo	18970	18580	-2.06	**0.0010**
Calgary	19289	19289	SOLDOUT	
Carolina	17560	15484	-11.83	**< 0.0001**
Chicago	21776	22623	3.89	0.5245
Colorado	15444	16296	5.51	0.0546
Columbus	14565	14698	0.91	0.8378
Dallas	17063	15421	-9.62	**0.0020**
Detroit[1]	20066	20066	SOLDOUT	
Edmonton	16839	16839	SOLDOUT	
Florida	16992	14525	-14.52	**< 0.0001**
Los Angeles	18179	19018	4.61	0.4685
Minnesota	18795	18506	-1.54	**0.0295**
Montreal	21273	21273	SOLDOUT	
Nashville	16974	16600	-2.20	**0.0117**
New Jersey	17114	15868	-7.28	0.2925
NY Islanders	13307	14741	10.77	0.2608
NY Rangers[2]	17200	18006	SOLDOUT	
Ottawa	19408	18109	-6.70	**0.0003**
Philadelphia	19787	19839	0.27	0.1515
Pittsburgh	18648	18618	-0.16	**0.0180**
San Jose	17562	17562	SOLDOUT	
St. Louis	17901	17498	-2.25	0.3766
Tampa Bay	19056	18612	-2.33	**0.0152**

Table 1 (continued)

Toronto	19426	19447	0.10	0.6859
Vancouver[3]	18910	18910	SOLDOUT	
Washington	18506	18506	SOLDOUT	
Winnipeg	15004	15004	SOLDOUT	

Source: Data from www.NHL.com , the official site of the NHL.
[1]Detroit hosted the Winter Classic in 2014, which was played outside with attendance over 105,000.
This data point was omitted from Detroit's home attendance figures.
[2]Madison Square Garden, home to the NY Rangers, was renovated between seasons, which increased
capacity from 17200 to 18006.
[3]Vancouver hosted the 2014 Heritage Classic, which was played outside with attendance over 54,000.
This data point was omitted from Vancouver's home attendance figures.

Table 2. Average Home Attendance by Team, 2011-12 and 2013-14 Seasons

Team	Average home attendance 2011-12	Average home attendance 2013-14	Percent change	*p*-value on difference
Anaheim	14760	16469	11.58	**< 0.0001**
Arizona	12421	13776	10.91	**0.0282**
Boston	17565	17565	SOLDOUT	
Buffalo	18550	18580	0.16	0.8434
Calgary	19289	19289	SOLDOUT	
Carolina	16043	15484	-3.48	0.1635
Chicago	21534	22623	5.06	0.2842
Colorado	15499	16296	5.14	**0.0399**
Columbus	14660	14698	0.26	0.9447
Dallas	14227	15421	8.39	0.0639
Detroit[1]	20066	20066	SOLDOUT	
Edmonton	16839	16839	SOLDOUT	
Florida	16628	14525	-12.65	**< 0.0001**
Los Angeles	17920	19018	6.12	0.2202
Minnesota	17773	18506	4.12	**0.0001**
Montreal	21273	21273	SOLDOUT	
Nashville	16691	16600	-0.54	0.5492
New Jersey	15397	15868	3.06	0.6103
NY Islanders	13191	14741	11.75	0.1193
NY Rangers[2]	17200	18006	SOLDOUT	
Ottawa	19356	18109	-6.44	**< 0.0001**
Philadelphia	19770	19839	0.35	0.2567
Pittsburgh	18566	18618	0.28	**< 0.0001**
San Jose	17562	17562	SOLDOUT	
St. Louis	18810	17498	-6.97	**< 0.0001**
Tampa Bay	18468	18612	0.78	0.4517

Table 2 (continued)

Toronto	19507	19447	-0.31	0.0949
Vancouver[3]	18890	18910	SOLDOUT	
Washington	18506	18506	SOLDOUT	
Winnipeg	15004	15004	SOLDOUT	

Source: Data from www.NHL.com, the official site of the NHL.
[1]See footnote 1 in Table 1.
[2]See footnote 2 in Table 1.
[3]See footnote 3 in Table 1.

Table 3. Average Home Attendance by Team, 2011-12 and 2012-13 Seasons

Team	Average home attendance 2011-12	Average home attendance 2012-13	Percent change	p-value on difference
Anaheim	14760	15888	7.64	**0.0100**
Arizona	12421	13924	12.10	**0.0475**
Boston	17565	17565	SOLDOUT	
Buffalo	18550	18970	2.26	**0.0175**
Calgary	19289	19289	SOLDOUT	
Carolina	16043	17560	9.46	**0.0008**
Chicago	21534	21776	1.12	**0.0067**
Colorado	15499	15445	-0.35	0.8994
Columbus	14660	14565	-0.65	0.8905
Dallas	14227	17063	19.93	**0.0002**
Detroit[1]	20066	20066	SOLDOUT	
Edmonton	16839	16839	SOLDOUT	
Florida	16628	16992	2.19	0.4250
Los Angeles	17920	18179	1.45	0.1650
Minnesota	17773	18795	5.75	**< 0.0001**
Montreal	21273	21273	SOLDOUT	
Nashville	16691	16974	1.70	0.0994
New Jersey	15397	17114	11.15	**< 0.0001**
NY Islanders	13191	13307	0.88	0.8486
NY Rangers[2]	17200	17200	SOLDOUT	
Ottawa	19356	19408	0.27	0.8011
Philadelphia	19770	19787	0.09	0.6391
Pittsburgh	18566	18648	0.44	**< 0.0001**
San Jose	17562	17562	SOLDOUT	
St. Louis	18810	17901	-4.83	**0.0025**
Tampa Bay	18468	19056	3.18	**0.0051**

Table 3 (continued)

Toronto	19507	19426	-0.42	**0.0174**
Vancouver[3]	18890	18910	SOLDOUT	
Washington	18506	18506	SOLDOUT	
Winnipeg	15004	15004	SOLDOUT	

Source: Data from www.NHL.com, the official site of the NHL.
[1]See footnote 1 in Table 1.
[2]See footnote 2 in Table 1.
[3]See footnote 3 in Table 1.

Reference

1. M. T. Nally and Paul M. Sommers, "Striking Back: The Baseball Fan Boycott of 1995." *Journal of Recreational Mathematics*, Vol. 29(3), 1998, pp. 184 – 188.

Footnote

1. A two-sample *t*-test compares two independent samples: 2012-13 game-by-game home attendance figures ($n_1 = 24$) with the larger number of 2013-14 game-by-game home attendance figures ($n_2 = 41$). A two-tailed *p*-value is reported in the last column of Table 1.

Mathematical Venery Revisited

Charles Ashbacher
cashbacher@yahoo.com

Abstract

In volume 15, number 3 of **Journal of Recreational Mathematics**, Charles W. Trigg defined mathematical venery to be the humorous act of assigning terms to groups of mathematical items. Some of his phrases were:

"A congruence of number theorists."

"An aberration of angle trisectors."

"A gardner of recreational mathematicians."

In this paper, some additional phrases are assigned to groups of mathematical items.

In the modern world, the term "venery" is used to refer to sexual indulgence. However, in the Old English world, venery referred to the naming of groups of animals all of the same type. It arose from the practice of group hunting. For example, a pride of lions, a collection of domestic turkeys is a rafter and a group of cats is a clowder.

In volume 15, number 3 of **Journal of Recreational Mathematics**, Charles W. Trigg lists some additional namings that he created in the area of mathematics. Some of them are: "A congruence of number theorists," "An aberration of angle trisectors" and "A gardner of recreational mathematicians."

Here are some additional assignments of names to groups of mathematical objects of my own creation.

A group of algebraists

A field of groups

A profligate of Fibonacci numbers

A rearrangement of combinatorists

A lo shu of magic squares

A random walk of investors

A compact of topologists

A tour of graph theorists

A ToE of string theorists

A manifold of Riemannian geometers

A ring of divisors

A domain of functions

A clique of graphs

A neighborhood of topologists

A galaxy of magic stars

A permutation of orders

A faithful representation of friends

A space of Euclidean geometers

A tree of computer scientists

Reference

C. W. Trigg, *Mathematical Venery*, **Journal of Recreational Mathematics**, 15(3). pp. 173, 1982-83.

Alphametics

Charles Ashbacher

1. The first problem is based on episode number 4 of the Star Trek Animated Series, "The Lorelei Signal." The 4 is the episode number and can be reused.

```
        4
      MEN
     WEAK
      DAY      WOMEN are greatest and solve in base 12
    SAVED
       BY
   _____

    WOMEN
```

2. This problem is based on episode 5 of the Star Trek Animated Series, "More Tribbles, More Troubles." The 5 is the episode number and can be reused.

```
         5
     JONES
      BACK     Solve in base 16. JONES is the greatest and make T &
      MORE     R perfect powers
    TRIBBLE
   _____

    TROUBLE
```

3. This problem is based on episode 62 of Star Trek: The Next Generation, "A Matter of Perspective."

```
        62
       DID
     RIKER   Solve in base 12 and maximize MURDER. The six and the two
    COMMIT   can be reused and the question mark is not part of the problem.
         A
   _____

    MURDER?
```

4. This problem is based on episode 27 of Star Trek: The Next Generation, "The Child." Solve in base 14 and since the alien grows so fast, we seek the largest ALIEN and since Troy and the alien are an odd pair, we seek an odd TROY.

```
     27
   TROY
  GIVES
  BIRTH
     TO
  _____
  ALIEN
```

5. This problem is based on episode 29 of "Star Trek: The Next Generation," "Elementary Dear Data." In this case, Geordi instructs the holodeck computer to create an adversary to Data as Sherlock Holmes that will be sufficiently challenging. The problem is to be solved in base 10, the 2 and the 9 can be reused, the question mark is not part of the problem and you are to find the greatest MATCH.

```
     29
    HAS
   DATA
    MET
    HIS
  _____
  MATCH?
```

6. This problem is based on episode 67 of "Star Trek: The Next Generation, Captain's Holiday." The 6 and 7 can be reused and it is to be solved in base 10.

```
      67
    VASH
   HARMS
  CHARMS
  _____
  PICARD
```

7. This problem is based on episode 88, "Star Trek: The Next Generation, Clues." The 8 can be reused and we seek the greatest CAUSE.

```
     88
   DATA
   LIES
    FOR
      A
   GOOD
   ─────
  CAUSE
```

8. This problem is based on episode 61, "Star Trek: The Next Generation, Deja Q." Solve in base 14, the six and one can be reused and we seek the maximum BORED.

```
      61
       Q
      IS
    NOWA
   BORED
   ─────
   HUMAN
```

9. This problem is based on episode 42, "Star Trek: The Next Generation, Q Who?" Solve in base 14, the four and two can be reused and we week the greatest PICARD.

```
      42
       Q
  STARTS
  BORGVS
  ──────
  PICARD
```

10. Mopan is a language in the Mayan family with approximately 12,000 native speakers in Guatamala and Belize. The following is a doubly true alphametic which is equivalent to $1 + 1 + 2 + 3 = 7$.

```
   JUN
   JUN    Solve in base 8, where JUN is odd, KA' is divisible by two and OX is
   KA'    divisible by three.
    OX
   ───
   WUK
```

11. The following is doubly-true in Cherokee and is to be solved in base 12.

```
    SAGWU      1
    SAGWU      1
    SAGWU      1
    SAGWU      1
    HISGI      5
  ─────────  ───
    SONELA     9
```

Furthermore, in the solution, HISGI is divisible by 5 and SONELA by 9.

12. The following is doubly-true in Basque and is to be solved in base 8.

```
    BAT        1
    BAT        1
    BAT        1
    BAT        1
    BAT        1
    BOST       5
  ─────────  ───
    HAMAR     10
```

13. This is an proverb that was revisited in "Star Trek: Into Darkness"

```
    ENEMY
       OFMY    In this case, your best FRIEND
    ENEMY
     ISMY
  ─────────
    FRIEND
```

41

THE NFL DRAFT: TWO MODELS FOR DRAFT EVALUATION

Andrew W. Jung
Paul M. Sommers

Middlebury College
psommers@middlebury.edu

Abstract

Using data on Day 1 NFL draft trades between 1985 and 2005, the authors examine how well (or poorly) two different draft trade-value charts explain career approximate values of previously drafted players. One chart, devised in the 1990s by then-Cowboys head coach Jimmy Johnson, enjoys widespread use (and appears at Pro-Football-Reference.com). The other, recently devised by Sommers, assigns more value to players picked after the very top of the draft. A series of t-tests and simple regression analysis show that the Sommers chart does a better job of predicting the eventual success of players exchanged on Day 1 of the annual NFL draft.

Introduction

The National Football League (NFL) Draft has been in existence since 1936. Currently the draft consists of seven rounds (the length of the draft since 1994). Under the NFL's reverse-order-of-finish draft, the weakest teams draft first and the winner of the Super Bowl drafts last. A team may give up a draft pick in a given round for one or more additional lower picks in the same or later rounds or for one or more picks in future drafts (or a combination of the two).

In the early 1990s, Mike McCoy, a Dallas Cowboys vice-president, and then-Cowboys head coach Jimmy Johnson devised a system to help evaluate NFL draft trades. Their trade-value draft chart is reproduced in Table 1. For each of the 224 picks (7 rounds, with each of 32 teams picking once per round), their chart assigns a value to each draft pick. The first overall pick is worth 3000 points, the second is worth 2600 points, and so forth until the last or 224th pick – worth only 2 points. This chart, reproduced at Pro-Football-Reference.com [1], enjoys widespread use to help teams approximate the market value of draft selection swaps. For example, a team might give up the 6th pick (1600 points) and get the 12th pick (1200 points) and 18th pick (400 points) in return.

In their research on overconfidence and market efficiency in the NFL, Massey and Thaler [2] conclude that top NFL draft picks are overvalued. They use data on draft-day trades for the years 1988 through 2004 to develop a model for predicting the market value of draft picks relative to the overall number one pick. Their analysis of draft data, player compensation, and performance measures (such as the number of games played, the number of games started, and the probability of making the Pro Bowl) leads them to the conclusion that teams put "too high a value on choosing early in the draft."

Sommers' [3] paper on winners and losers in the NFL draft between 2002 and 2014 endeavors to value each draft pick based on a player's approximate value (hereafter AV), a comprehensive metric (reported for each NFL player at Pro-Football-Reference.com) to assess player worth at any position in any given year. Sommers relates draft pick number to the average career AVs of all players drafted by the NFL between 2002 and 2014. The Sommers trade-value draft chart[1] is reproduced in Table 2 and recognizes that NFL teams have been putting "too high a value" on players near the top of the draft. For example, a team might give up the 6th pick (2106 points) and get the 12th pick (1759 points) and, unlike the Johnson chart, the 203rd pick (348 points) in return. The Sommers model indicates that the tenth pick is worth (1851/3000 or) about 62 percent of the first overall pick and the 20th is worth (1504/3000 or) a little over 50 percent of the first overall pick. By comparison, the Johnson value chart indicates that the tenth pick is worth (1300/3000 or) about 43 percent of the first overall pick and the 20th is worth (850/3000 or) 28 percent of the first overall pick. While the last (or 224th) pick on Johnson's value chart receives only 2 points, Sommers' model awards the last pick 298 points – on the same scale (where the first overall pick receives 3000 points). The drop in value using the Johnson chart from the first pick to the tenth is roughly 57 percent, and another 55 percent drop from there to the end of the first round. By comparison, the drop in value using the Sommers model from the first pick to the tenth is roughly 38 percent, and another 31 percent drop from there to the end of the first round. Summarizing, the draft-pick value declines much more steeply in the Johnson model than in the Sommers model.

In the next section, we address the key question — which model (Johnson or Sommers) better predicts "good deals" from "bad deals"?

The Data

We use data on all draft selection swaps on Day 1 (sometimes referred to as Round 1) of the NFL draft for the years 1985 through 2005.[2] Over this 21-year period we observe 95 draft-day trades involving draft picks from only the current year (78) or from the current and future years (17). We exclude trades that involve NFL players. For each team involved in a trade, we recorded the acquired player's name, his overall pick number, and his chart value according to Johnson and Sommers. For each drafted player, we also collected data on the drafted player's eventual career AV [4].[3] The chart values of future picks were discounted at a 5 percent rate. (For example, the Johnson chart value of the overall 6th pick next year would be 1600/1.05 or 1523.8 points.) We then added up each team's total point valuation. Finally, for each team, we added up the players' NFL career AVs. The data on all 95 trades appear in Table 3.

For each trade and each model (Johnson or Sommers), we derived a disparity measure, D. If the Johnson model, for example, predicts that Team A will end up with more point value than Team B, then the disparity value is equal to the cumulative career AV of all picks acquired by Team A *minus* the cumulative career AV of all picks acquired by Team B. That is, if the Johnson model correctly predicts the ith trade, then the disparity measure, D_i, will be positive. If, however, the Johnson model predicts one team has a point advantage, but in fact the *other* team ends up acquiring players with a higher cumulative total of career AVs, then the disparity measure will be negative. We investigate whether the average value of $D_{Johnson}$ (or $D_{Sommers}$) is equal to zero against the one-tailed alternative that the average value is greater than zero. If we cannot reject the null hypothesis that the average value of $D_{Johnson}$ (or $D_{Sommers}$) is equal to zero, then the chart's ability to distinguish good trades from bad is no better than a coin flip.

Finally, for each model, we regress the difference between the total career AVs of Team A's draft picks minus the total career AVs of Team B's draft picks (*CareerAV_Difference*) against the corresponding difference in the total point values of draft picks (*Point_difference*), as follows:

$$CareerAV\_difference_i = \beta_0 + \beta_1 Point\_difference_i + \varepsilon_i$$

where ε_i are independent errors. If the slope coefficient, β_1, is not discernible from zero, then variation in the difference between Team A's points less Team B's points tells us nothing about the variation in the team differences in career AVs. If, however, the slope coefficient is significantly greater than zero, then the model provides evidence of a direct relationship between the two variables. That is, if the difference between the point values of Team A's picks less Team B's picks is positive (negative), then the difference between the career AVs of drafted players acquired by Team A less the career AVs of those players acquired by Team B will also be positive (negative).

The Results

For the 95 draft trades, the average Johnson model disparity, $\overline{D}_{Johnson}$, is 4.23 with a *p*-value of .278. In other words, the average disparity using the Johnson chart is not discernible from zero. The average Sommers model disparity, $\overline{D}_{Sommers}$, is 16.25 with a *p*-value of .011. In other words, we can reject the null hypothesis, H_0: average of $D_{Sommers} = 0$, in favor of the one-tailed (greater than) alternative, H_A: average of $D_{Sommers} > 0$. That is, over the 21-year period, the Sommers chart was more often than not correct in predicting the two teams' difference in career AVs.[4]

When the difference in Team A's less Team B's career AVs was regressed against the difference in Team A's less Team B's Johnson point totals, the results were (*t*-values in parentheses):

(1)   CareerAV_difference = 6.642 + .0059 Johnson_point_difference
                        (0.91)   (0.24)
                                                $R^2 = .0006$

The corresponding regression results using Sommers point totals were:

(2)   CareerAV_difference = -5.187 + .0195 Sommers_point_difference
                        (-0.60)   (2.36)
                                                $R^2 = .0564$

Using the Johnson chart, equation (1) indicates that knowledge of the difference in point totals ascribed to the draft picks of Teams A and B sheds no light on the ultimate career AVs of players drafted by the two teams. Equation (2) indicates that if the point total of players drafted by Team A exceeds that of players drafted by Team B, then the ultimate total career AVs of the players acquired by Team A will exceed the corresponding total career AVs of the players acquired by Team B. In other words, the slope coefficient in equation (1) is not discernibly different from zero ($p = .405$); the slope coefficient in equation (2) *is* significantly greater than zero ($p = .011$).

Concluding Remarks

An NFL Draft valuation model that can consistently predict draft pick trade outcomes in terms of actual career AV would be highly valuable to general managers and coaching staffs of NFL teams.

The Johnson chart has enjoyed widespread use. But, if players near the top of the draft are indeed overvalued relative to players that come afterwards, then an alternate chart that assigns more value to players drafted in lower rounds may align better with the career AVs of these drafted players.

The results presented here show that the alternate chart generated by the Sommers model is not only better at predicting the winner of a draft trade, but that over a 21-year period there's a strong direct relationship between the Sommers point difference (Team A's total minus Team B's total) and the actual career AV difference of these acquired draft picks (again, Team A's total minus Team B's total). There were 52 (of 95) trades when the Johnson chart predicted an equal exchange (that is, the difference between the total point value of Team A's draft picks less Team B's picks was less than 10 percent) and the Sommers model predicted that there was *more than* a 10 percent difference in total point value. Of these 52 trades, 47 actually turned out to be uneven (using career AVs), as Sommers predicted. This is most likely the case because when the Johnson trade-value chart is used (which appears to be quite often) the total point values of the draft picks of the two teams are roughly the same.[5]

A disparity measure based on the career AVs of draft picks and the predictive accuracy of the trade value chart was not, on average, discernably different from zero for the Johnson model. The average disparity was discernably *greater than* zero for the Sommers model. Finally, a simple regression model shows that the Sommers value chart does a better job of explaining variation in the draft picks' eventual career AVs.

Table 1
Jimmy Johnson NFL Draft Trade Value Chart

Pick number	Round						
	1	2	3	4	5	6	7
1	3000	580	265	112	43	27	14.2
2	2600	560	260	108	42	26.6	13.8
3	2200	550	255	104	41	26.2	13.4
4	1800	540	250	100	40	25.8	13
5	1700	530	245	96	39.5	25.4	12.6
6	1600	520	240	92	39	25	12.2
7	1500	510	235	88	38.5	24.6	11.8
8	1400	500	230	86	38	24.2	11.4
9	1350	490	225	84	37.5	23.8	11
10	1300	480	220	82	37	23.4	10.6
11	1250	470	215	80	36.5	23	10.2
12	1200	460	210	78	36	22.6	9.8
13	1150	450	205	76	35.5	22.2	9.4
14	1100	440	200	74	35	21.8	9
15	1050	430	195	72	34.5	21.4	8.6
16	1000	420	190	70	34	21	8.2
17	950	410	185	68	33.5	20.6	7.8
18	900	400	180	66	33	20.2	7.4
19	875	390	175	64	32.6	19.8	7
20	850	380	170	62	32.2	19.4	6.6
21	800	370	165	60	31.8	19	6.2
22	780	360	160	58	31.4	18.6	5.8
23	760	350	155	56	31	18.2	5.4
24	740	340	150	54	30.6	17.8	5
25	720	330	145	52	30.2	17.4	4.6
26	700	320	140	50	29.8	17	4.2
27	680	310	136	49	29.4	16.6	3.8
28	660	300	132	48	29	16.2	3.4

Table 1 (continued)

29	640	292	128	47	28.6	15.8	3
30	620	284	124	46	28.2	15.4	2.6
31	600	276	120	45	27.8	15	2.3
32	590	270	116	44	27.4	14.6	2

Source: http://www.pro-football-reference.com/draft/draft_trade_value.htm

Table 2
Sommers NFL Draft Trade Value Chart

				Round			
Pick number	1	2	3	4	5	6	7
1	3000	1254	916	716	574	463	373
2	2654	1240	908	711	570	460	370
3	2452	1225	901	706	566	457	368
4	2308	1211	894	701	562	454	365
5	2197	1197	886	696	559	451	363
6	2106	1184	879	691	555	448	360
7	2029	1171	872	686	551	445	357
8	1962	1158	865	681	548	442	355
9	1903	1146	858	677	544	439	352
10	1851	1134	851	672	540	436	350
11	1803	1122	845	667	537	433	348
12	1759	1111	838	663	533	430	345
13	1720	1100	831	658	529	427	343
14	1683	1089	825	653	526	425	340
15	1648	1078	819	649	522	422	338
16	1616	1067	812	644	519	419	335
17	1586	1057	806	640	516	416	333
18	1557	1047	800	636	512	413	331
19	1530	1037	794	631	509	410	328
20	1504	1027	788	627	505	408	326
21	1480	1018	782	623	502	405	324
22	1457	1009	776	618	499	402	321
23	1435	999	771	614	495	399	319
24	1413	990	765	610	492	397	317
25	1393	982	759	606	489	394	314
26	1373	973	754	602	485	391	312
27	1355	964	748	598	482	389	310
28	1337	956	743	594	479	386	307

Table 2 (continued)

29	1319	948	737	590	476	383	305
30	1302	940	732	586	473	381	303
31	1286	932	727	582	470	378	301
32	1270	924	721	578	466	375	298

Table 3

NFL Draft Trades and Trade Values, Round 1, 1985-2005

Year	Team A	Team B	Johnson Trade Value Team A	Team B	Sommers Trade Value Team A	Team B	Career AV Team A	Team B
1985	Vikings	Falcons	2100	2600	3264	2654	157	70
1985	Patriots	49ers	1170	1215	3115	2461	28	252
1985	Oilers	Vikings	2820	2600	3754	2654	148	70
1985	Bills	Packers	1580	1661.9	2817	2779.5	57	133
1985	Browns	Bills	3000	2756.8	3000	4961.9	80	172
1986	49ers	Cowboys	891	900	2070	1557	102	38
1986	49ers	Bills	980	852	2309	1704	64	101
1986	Saints	Colts	1900	1800	3062	2308	192	52
1986	Vikings	Chargers	1560	1660	3094	2870	92	120
1987	Bills	Oilers	1940	2200	3173	2452	129	12
1987	Redskins	49ers	450	687.7	1100	1531.7	3	6
1987	Dolphins	Vikings	1044	1100	2194	1683	15	8
1988	Lions	Chiefs	2840	2600	3771	2654	146	120
1989	Bengals	Falcons	697	680	2216	1355	7	11
1989	Bears	Dolphins	805	720	2127	1393	117	50
1989	Broncos	Browns	1321	1150	3399	1720	144	77
1990	Steelers	Cowboys	985	950	2286	1586	46	170
1990	Patriots	Seahawks	3061.4	2840	5399.9	3771	120	181
1991	Cowboys	Redskins	890	950	2066	1586	48	13
1991	Eagles	Packers	1400	1779.8	1962	3040.5	41	63
1991	Oilers	Patriots	756	950	2033	1586	28	13
1991	Cowboys	Patriots	1024	1100	2239	1683	13	28
1992	Falcons	Cowboys	961	1004	2211	2196	11	43
1992	Bengals	Redskins	2430	2120	4231	3281	87	24
1992	Cowboys	Patriots	1491	1385	3408	2592	107	33
1992	Patriots	Falcons	1401	1400	3300	1962	11	94
1993	Saints	49ers	850	885	1504	2179	16	143
1993	49ers	Cardinals	912	900	2131	1557	33	11
1993	Cowboys	Packers	930.2	710	3154	1963	105	88
1993	Oilers	Eagles	1150	1090	1720	2375	94	49
1993	Browns	Broncos	1275	1250	2477	1803	78	36
1994	49ers	Cowboys	944	764.6	2277	1749	25	25
1994	Dolphins	Packers	995	1000	2263	1616	83	38
1994	Saints	Jets	1184.5	1200	2242	1759	66	95
1994	Browns	Eagles	640	804.8	1319	2084.7	29	42
1994	Rams	49ers	1490	1500	3339	2029	113	118
1994	Rams	Colts	1675	1700	2823	2197	118	4

Table 3

NFL Draft Trades and Trade Values, Round 1, 1985-2005

(continued)

Year	Team A	Team B	Johnson Trade Value		Sommers Trade Value		Career AV	
			Team A	Team B	Team A	Team B	Team A	Team B
1995	Cowboys	Buccaneers	766	660	2078	1337	2	194
1995	Panthers	Packers	796.2	877.2	1843	2613	53	42
1995	Jaguars	Chiefs	875	815.8	1530	3166.5	52	8
1995	Chargers	Panthers	768	640	2652	1319	38	47
1995	Panthers	Bengals	2240	3000	3408	3000	122	13
1995	Vikings	Falcons	1750	450	2961	1100	39	44
1995	Browns	49ers	1466.7	1300	3955.6	1851	232	47
1995	Eagles	Buccaneers	1730	1946	2894	3813	58	166
1996	Bears	Rams	1150	1086	1720	2703	82	91
1996	Oilers	Raiders	1446	1350	3311	1903	20	27
1996	Seahawks	Lions	936	950	2228	1586	88	11
1996	Redskins	Cowboys	620	785	1302	2098	0	54
1997	Oilers	Chiefs	1117	1224	3337	2373	144	149
1997	Buccaneers	Seahawks	1476	1600	2691	2106	165	127
1997	Rams	Jets	3000	1955.6	3000	4036	128	190
1997	Chargers	Buccaneers	490	1000	1146	1616	8	22
1997	Falcons	Seahawks	2080	2476	4529	3384	61	121
1997	Eagles	Cowboys	978	780	2712.1	1457	9	7
1997	Jets	Buccaneers	1486	1600	2643	2106	139	127
1998	Buccaneers	Raiders	760	870	1435	2204	0	41
1998	Dolphins	Packers	940	875	2275	1530	41	84
1999	Dolphins	49ers	719	740	1910	1413	17	1
1999	Dolphins	Lions	785	680	2576	1355	31	14
1999	Seahawks	Patriots	1045	950	2682	1586	42	74
1999	Seahawks	Cowboys	816	850	1990	1504	31	40
1999	Bears	Redskins	1699.1	1500	4559.3	2029	62	151
1999	Redskins	Saints	2106.3	1700	6187.6	2197	120	90
2000	49ers	Jets	1420	1200	2683	1759	117	95
2000	49ers	Redskins	1940	2200	3172	2452	120	76
2000	Broncos	Ravens	1500	1300	2748	1851	93	36
2001	Colts	Giants	770.2	780	2423	1457	166	55
2001	Seahawks	49ers	1532.6	1515	3006	2407	66	76
2001	Bills	Buccaneers	1190	1100	2517	1683	82	32
2001	Steelers	Jets	966	1000	2584	1616	103	93
2002	Titans	Giants	1124	1100	2301	1683	62	59

Table 3

NFL Draft Trades and Trade Values, Round 1, 1985-2005

(continued)

Year	Team A	Team B	Johnson Trade Value		Sommers Trade Value		Career AV	
			Team A	Team B	Team A	Team B	Team A	Team B
2002	Falcons	Raiders	928.2	950	2030	1586	24	40
2002	Redskins	Raiders	945	900	2239	1557	79	23
2002	Seahawks	Packers	960	879	2293	1983	24	108
2002	Redskins	Patriots	708	800	2289	1480	16	22
2002	Cowboys	Chiefs	1631.2	1600	3179.4	2106	72	28
2003	Bears	Patriots	1114.2	1150	2056	1720	8	50
2003	Cardinals	Saints	2210	2222	4152	3994	235	43
2003	Bears	Jets	1992	1800	3804	2308	98	38
2003	Chargers	Eagles	904	1050	2242	1648	13	3
2003	Chiefs	Steelers	823.4	1000	2453	1616	64	115
2003	Patriots	Ravens	1251.9	875	2555.5	1530	110	18
2004	Titans	Texans	860	707.8	3256	1825	129	53
2004	Vikings	Dolphins	906	875	2118	1530	53	51
2004	49ers	Eagles	980	1000	2310	1616	86	40
2004	Chiefs	Lions	655	620	2372.8	1302	20	29
2004	Bengals	Rams	749	740	1971	1413	34	92
2004	Colts	Falcons	812	780	2660	2073	37	125
2004	49ers	Panthers	645	660	1868	1337	1	55
2004	Cowboys	Bills	1313.5	780	3073.4	1457	85	20
2004	Lions	Browns	2030	1600	3226	2106	58	43
2005	Seahawks	Raiders	784	760	2050	1435	56	23
2005	Texans	Saints	1247.6	1150	2480.8	1720	80	54
2005	Broncos	Redskins	1006.2	720	2810.4	1393	161	53

References

1. Draft trade value chart. (n.d.). Retrieved from www.pro-football-reference.com/draft/draft_trade_value.htm

2. C. Massey and R. Thaler, "Overconfidence vs. Market Efficiency in the National Football League." National Bureau of Economic Research, Working Paper 1270, April 2005.

3. P.M. Sommers, "The NFL Draft, 2002 – 2014: Winners, Losers, and a New Draft Trade Value Chart." *Topics in Recreational Mathematics*, volume 8, forthcoming.

4. Day 1 draft trades. (n.d.). Retrieved from www.prosportstransactions.com/football/DraftTrades/Years/ .

5. Players' career AV. (n.d.). Retrieved from www.pro-football-reference.com/players/ .

Footnotes

1. The value of a draft pick (i.e., the player's career AV) as a function of the pick number was estimated by Sommers as follows (*t*-values in parentheses):

   $$CareerAV = 53.6728 - 8.9315\ ln(pick\ number)$$
   $$(37.87) \quad (-28.54)$$
   $$R^2 = .79$$

   The overall first pick in the draft has (like Johnson's standard) a value of 3000 points. Accordingly, the Sommers value of any other pick number would be:

   $$[53.6728 - 8.9315\ ln(pick\ number)] \times (3000/53.6728)$$

2. The data set was compiled from Pro Sports Transactions [4].

3. Only ten of the 359 players whose career AVs we recorded were still playing in the NFL as of 2015 [Calvin Pace, traded in 2003 by the Saints to the Cardinals; Anquan Boldin, 2003, Saints to Cardinals; Vince Wilfork, 2003, Ravens to Patriots; Steven Jackson, 2004, Bengals to Rams; Randy Starks, 2004, Texans to Titans; Jason Babin, 2004, Titans to Texans; Matt Schaub, 2004, Colts to Falcons; Eric Winston, 2005, Saints to Texans; Manny Lawson, 2005, Redskins to Broncos; and Brandon Marshall, 2005, Redskins to Broncos].

4. When the disparity measure (that is, the difference between career AVs of the teams' draft picks, which is positive if the chart correctly predicts the team that got the better deal and negative if not) is expressed as a percentage of the average career AVs of the two teams' draft picks, the *t*-value on this adjusted disparity measure is 1.105 for the Johnson model ($p = .136$) and 2.717 ($p = .004$) for the Sommers model. Once again, the Sommers model appears to be the better predictor of winners versus losers in draft trades.

5. For all 95 trades, the average point difference (Team A's total minus Team B's total) using the Johnson chart was not discernible from zero ($p = .112$).

Wordplay and Humor

Rachel Pollari

Charles Ashbacher

Written by Rachel Pollari

The atmosphere is discharging felines and canines!

Allow snoring pooches to remain prone.

Emancipate the mouser from the haversack.

You ascertained it directly from the equine's maw.

As the Corvus maneuvers while airborn…

You're yowling in proximity of the incorrect sapling.

Feline acquired your mouth's red fleshy muscular organ?

They are plunging downward like insects of the order Diptera!

Maintain your equines!

Primate view, primate accomplish.

Seize the uncastrated male bovine by the rack!

Consecrated dairy mammal!

Written by Charles Ashbacher

Famous movie quotes altered to reflect mathematical themes

"Frankly my dear, I don't give a delta."

From "Gone With the Wind"

"You would prefer another function? A continuous function? Then name the system!"

From "Star Wars Episode IV: A New Hope"

Trimorphic Numbers Revisited

Charles Ashbacher
cashbacher@yahoo.com

Abstract
In volume 17, number 4 of "Journal of Recreational Mathematics," Charles W. Trigg defined a **trimorphic number** to be a triagonal number $T(n) = [n(n + 1)] / 2$ that terminates in n. For example, $T(25) = 325$ and $T(625) = 195625$ and
$T(9376) = 43959376$. The purpose of this paper is to report the results of a greater search for trimorphic numbers.

Note: There is another set of numbers that are defined as trimorphic. There is the cubes that terminate with n. In other words n^3 terminates with n.

In volume 17, number 4 of "Journal of Recreational Mathematics," Charles W. Trigg started with the triagonal numbers defined by the formula $T(n) = [n(n + 1)] / 2$ and defined the **trimorphic** numbers to be triagonal numbers that terminate with the index of the triagonal number.

Trigg identified six base-ten trimorphic numbers less than 10^5 and they are: $T(1) = 1$, $T(5) = 15$, $T(25) = 325$, $T(625) = 195625$, $T(9376) = 43959376$ and $T(90625) = 4106490625$.

I recently encountered this paper when I was reading through old issues of **JRM** and wondered if any additional work had been done in this area. Naturally, I did a Google search but there was no mention of the trimorphic numbers as defined by Trigg. There is another set of numbers called trimorphic that are the set of cubes n^3 that terminate with n. Therefore, using the BigInteger class in the progamming language Java, I conducted a more extensive search for additional octamorphic numbers in base ten.

The search was extended up to 10^9 and additional trimorphic numbers were found. They appear in table 1.

Table 1

n	T(n)
890625	396606890625
7109376	25271617109376
12890625	83084112890625
212890625	22661209212890625

The most obvious point of interest is the values of the last three and four digits, which are repeats of the digits of the larger two of the trimorphic numbers found by Trigg. This leads to two additional unresolved questions.

1. Are there additional trimorphic numbers?

2. If there are more trimorphic numbers, do they have the same trailing digits as those in table 1 or are they different?

The trailing digits of the values of n of the known trimorphic numbers indicate the potential for an infinite family.

Reference

1. C. W. Trigg, *Trimorphic Numbers*, **Journal of Recreational Mathematics**, 17(4). pp. 282, 1984-85.

Which Gender is Happier in the United States?

Charles Ashbacher
cashbacher@yahoo.com

Abstract

Given any integer n > 0, the sum of the squares of the digits can be computed. As an example, for the number 12, the value is $1^2 + 2^2 = 5$. This process can be iterated and there are two possible outcomes, the process eventually yields 1 or it enters into a cycle. For example, the process for the number 13 is

$1^2 + 3^2 = 1 + 9 = 10$ $1^2 + 0^2 = 1$.

For the number 4, the process is

$4^2 = 16$, $1^1 + 6^2 = 37$, $3^2 + 7^2 = 58$, $5^2 + 8^2 = 89$, $8^2 + 9^2 = 145$, $1^2 + 4^2 + 5^2 = 42$, $4^2 + 2^2 = 20$,

$2^2 + 0^2 = 4$. This is a cycle that will repeat indefinitely. When the process terminates at 1, the number is said to be **happy**[1].

The letters of the alphabet can be assigned numeric values in the following way:

a = 1, b = 2, c = 3, . . . , z = 26.

Using this coding, every word can be converted into a positive integer and that number can be tested to determine if it is happy. If it is, then the word is said to be a **happy word**[2]. If the word is a name, then it is said to be a **happy name**.

The U. S. Social Security Administration maintains a list of the 100 most popular names over the last century (1916-2015)[3]. That list is separated into the 100 most popular male and 100 most popular female names. Each of those lists was examined for the presence of happy names and this paper is a report on those results.

Introduction

Given any integer n > 0, the sum of the squares of the digits can be computed. As an example, for the number 12, the value is $1^2 + 2^2 = 5$. This process can be iterated and there are two possible outcomes, the process eventually yields 1 or it enters into a cycle. For example, the process for the number 13 is

$1^2 + 3^2 = 1 + 9 = 10$ $1^2 + 0^2 = 1$.

For the number 4, the process is

$4^2 = 16$, $1^1 + 6^2 = 37$, $3^2 + 7^2 = 58$, $5^2 + 8^2 = 89$, $8^2 + 9^2 = 145$, $1^2 + 4^2 + 5^2 = 42$, $4^2 + 2^2 = 20$,

$2^2 + 0^2 = 4$. This is a cycle that will repeat indefinitely. When the process terminates at 1, the number is said to be **happy**[1].

The letters of the alphabet can be assigned numeric values in the following way:

a = 1, b = 2, c = 3, . . . , z = 26.

Using this coding, every word can be converted into a positive integer and that number can be tested to determine if it is happy. If it is, then the word is said to be a **happy word**[2]. If the word is a name, then it is said to be a **happy name**.

For example,

emma, e = 5, m = 13, a = 1 for a total of 32. $3^2 + 2^2 = 9 + 4 = 13$. $1^2 + 3^2 = 1 + 9 = 10$. $1^2 + 0^2 = 1$, so emma is a happy name.

Data Acquisition

The United States Social Security Administration maintains a database of the 100 most popular names in the United States over the years 1916-2015. Two separate lists of 100 male and 100 female names are kept. To simplify the process, all names were converted to all lowercase. If the same name appeared in slightly different form, both were unaltered. For example, Sara and Sarah both appear in the list.

Data Analysis

A program was written in Java to read the names from the lists and count the number of names that were happy. Of the 100 most common male names, 23 were happy and of the 100 most common female names, 14 were happy. Therefore, using this criteria, males in the United States are happier than females.

References

1. http://mathworld.wolfram.com/HappyNumber.html

2. **Number Treasury^3: Investigations, Facts and Conjectures about More than 100 Number Families**, Margaret J. Kenney and Stanley J. Bezuszka, World Scientific Press, 2015.

3. https://www.ssa.gov/oact/babynames/decades/century.html

Fun With Words

Charles Ashbacher

cashbacher@yahoo.com

In the video accessed from the **Math's Believe it or Not!** link on Facebook

https://www.youtube.com/watch?v=z6jMU-AwX34&feature=youtu.be

math comedian Matt Parker lists some features of the number 2017 on the advent of the new year. One of the characteristics is that when the number is written out in English there are 11 distinct letters, which are

{ 'a', 'd', 'e', 'h', 'n', 'o', 's', 't', 'u', 'v', 'w' }.

He then notes that it has no letters in common with the phrase "prickly fig."

After a moment's thought, I wondered, "What is the longest English word that has no letter in common with 2017 when written in English?"

To examine this, I needed a list of words, so I went to the site

http://www-01.sil.org/linguistics/wordlists/english/

and downloaded the file wordsEn.txt that contains 109,582 English words. Once the file was created I wrote a Java program that would read the words in the file and determine which ones did not have a letter in common with 2017. My interest was in determining what the longest word was that did not share a letter with 2017 and the program easily carried out that task.

The four longest such words were "bicyclic", "cyclicly", "frizzily" and "gimmickry."

This short paper by Rudolph Ondrejka appeared in Issue Number 6 of "Recreational Mathematics Magazine." The first sentence refers to the following short note that appeared in RMM issue number 1.

2. SOME NUMBER TOUGHIES

What two integers, neither containing any zeros, when multiplied together equal exactly 1,000,000,000? If that was too easy (or too hard) what two integers, also containing no zeros, multiply to give exactly one quintillion? (i.e. 1,000,000,000,000,000,000) H. V. Gosling

NON-ZERO FACTORS OF 10^n

by Rudolph Ondrejka

Since H. V. Gosling's problem about pairs of factors of 1 billion and 1 quintillion (10^9 and 10^{18}) which contain no zeroes (RMM No. 1, February 1961, page 44), T. H. Engel found the next highest power of 10 which has two factors with non-zero digits: 10^{33} (RMM No. 3, June 1961, page 58).

Naturally, one wonders if there are any other powers of 10, either greater or less than given above, with the aforementioned property. Factoring 10^n as $(2^n)(5^n)$ we find that there are only a limited number of powers of 2 which have nonzero digits. Matching these powers of 2 with powers of 5 which have non-zero digits quickly reduces the total number of pairs of factors.

The following are the only known powers of 10, up to 10^{5000}, which have the property that the two factors shown are composed of non-zero digits:

$10 = (2)(5)$
$10^2 = (2^2)(5^2) = (4)(25)$
$10^3 = (2^3)(5^3) = (8)(125)$
$10^4 = (2^4)(5^4) = (16)(625)$
$10^5 = (2^5)(5^5) = (32)(3125)$
$10^6 = (2^6)(5^6) = (64)(15625)$
$10^7 = (2^7)(5^7) = (128)(78125)$
$10^9 = (2^9)(5^9) = (512)(1953125)$
$10^{18} = (2^{18})(5^{18}) = (262144)(3814697265625)$
$10^{33} = (2^{33})(5^{33}) = (8589934592)(116415321826934814453125)$

The next power of 10, if one should exist with this property, would be greater than 10^{5000}. When you consider that the largest value of 2^n with non-zero digits is 2^{86} (and this has been checked out to 2^{5000}, though it is conceivable that a larger power of 2 might be composed of non-zero digits) then it becomes apparent that the probability of a power of 10, greater than 10^{33}, having two factors with non-zero digits is practically negligible.

Addendum by Charles Ashbacher

When I read this short piece, one of my first thoughts dealt with the powers of the duodecimal base 12 and how many equal powers of $2^2 = 4$ and 3 there are such that both powers have no non-zero digits. Since one of the base factors is a power of 2, there are fewer opportunities for the absence of zeros in those products, however since it is possible that there are more powers of three with no zero digits, this is not conclusive.

The first follow the sequence of the powers of 10

$12 = (3)(4)$
$12^2 = (3^2)(4^2) = (9)(16)$
$12^3 = (3^3)(4^3) = (27)(64)$
$12^4 = (3^4)(4^4) = (81)(256)$

However, since $4^5 = 1024$, we have the first break from the pattern of the powers of 10. Continuing on, we have other instances where there is the simultaneous lack of zeros in the powers when $n = 7$, $n = 8$, $n = 9$, $n = 12$ and $n = 14$.

$12^7 = (3^7)(4^7) = (2187)(16384)$
$12^8 = (3^8)(4^8) = (6561)(65536)$
$12^9 = (3^9)(4^9) = (19683)(262144)$
$12^{12} = (3^{12})(4^{12}) = (531441)(16777216)$
$12^{14} = (3^{14})(4^{14}) = (4782969)(268435456)$

There are no other nonzero factors up to $n = 5000$.

Abstracts of the papers in "Journal of Recreational Mathematics" Volume 1, Numbers 2, 3 and 4, 1968

Charles Ashbacher
cashbacher@yahoo.com

Given the elapsed time since these papers appeared and that there were no abstracts with the originals, all of the items in this list were written by Charles Ashbacher. This is yet another iteration in his ongoing project to make the material published in "Journal of Recreational Mathematics" public and accessible.

Number 2

"Compound Games With Counters," by Cedric A. B. Smith

Abstract
Many different games can be defined where there is one or more pile of counters and two players alternate removing a nonzero number of counters. The general rule is that the number of counters removed must be less than a fixed number M. Winners of games of this type can either be declared the winner or loser if they remove the last counter.

This paper contains an analysis of several games of this type and some of them are Nim or a variation. When there are multiple piles, the games can be played in either a conjunctive or disjunctive mode. In a conjunctive mode, the players must remove at least one counter from every pile in a move and in a disjunctive mode, counters are removed from only one pile in a move. Another variant has the player making a move in at least one pile but not in all.

"Curiosa on 1968," by Leon Bankoff

Abstract
This short paper contains fifteen ways of expressing 1968 cubed as the sum of three different cubes of positive integers as well as two basic identities using the Fibonacci numbers.

"Some Approximate Dissections," by Harry Lindgren

Abstract
In this paper, polygons are dissected into two or more pieces, where the pieces are very close to identical. For example, in the first dissection, the two pieces are in the ratio of 99:101. The figure dissected is in nearly all cases the regular pentagon. The pieces that it is dissected into are the equilateral triangle, square, the regular hexagon and Greek and Latin crosses. In nearly all cases the error is so small as to be capable of fooling a viewer if drawn correctly. The non-pentagonal figures dissected are the hexagon and octagon.

"Patterns in Primes," by Leslie E. Card

Abstract

There are many different patterns that one can find in the prime numbers. In this case, prime numbers that are reversible, cyclic and can be used to create squares and cubes are presented. Another interesting operation is to start with a 5-digit prime and add digits one at a time to form a continuous series of overlapping 5-digit primes.

"A Prime Number Sieve," by Douglas A. Engel

Abstract

Lines with slopes of 1/1, 1/2, 1/3, . . . starting at (0,0) are drawn in the first quadrant of the coordinate plane. Drawing vertical lines down from the lattice points on the line with slope 1/1 and checking for points of intersection of the other lines will reveal primes and composites.

"A Digital Bracelet for 1968," by Charles W. Trigg

Abstract

A digital bracelet is constructed by starting with four digits, computing the unit digit of their sum, affixing that to the end, then computing the unit digit of the sum of the preceding four digits and then repeating the process. Given the limited number of four digit numbers, the operation will eventually repeat and that cycle for starting with 1968 is given. Other features pointed out are the patterns that emerge when the elements of the cycle are placed in rows in a certain way.

"'Hit-and-Run' on a Graph," by Jorg Nievergelt and Steve Chase

Abstract

Shannon's switching game begins with a graph with two distinguished nodes. There are two players that are called "Cut" and "Short" respectively. For each play, Cut deletes a node connection while Short claims a connection as immune from cutting. Short wins if a path between the two distinguished nodes is preserved at the end while Cut wins if there is no such path.

Hit-and-Run is a similar game played on a 4×4 board by two players called H and V. The goal of H is to add node connections so that there is a horizontal path between two nodes on the vertical sides and the goal of V is to create a vertical path between the nodes on the horizontal sides. This paper is a report on a computer program that was written to play the game and insights that were generated.

"A 1968 Magic Square Composed of Leap Years," by Leon Bankoff

Abstract

This is a four-by-four magic square composed of leap years where the magic sum is 1968.

"Integers of the Form $N^3 + M^5$," by Charles W. Trigg

Abstract

This is a complete list of all integers P < 10,000 which are the sum of positive cube and positive fifth powers.

"A Special Square Array of the Nine Digits," by Charles W. Trigg

Abstract

The digits one through 9 are put into a three-by-three array that has a set of interesting properties when rotated.

"Square Solitaire and Variations," by Donald C. Cross

Abstract

Square solitaire is a game where fifteen pegs are placed on a four-by-four board with sixteen holes. A move is when one peg jumps another to an empty space and the jumped peg is removed. The goal is to complete a set of moves so that only one peg remains.
Solutions to the four-by-four game are given as well as a solution for a five-by-five, a seven-by-seven and an eight-by-eight game.

"A Fundamental Dissection Puzzle," by Kobon Fujimura

Abstract

The dissections examined in this paper are those of a square board that is segmented into squares of equal size. All dissection lines must follow the borders of squares. Four-by-four and five-by-five boards are dissected into identical parts and when there is an odd number of squares, the central one is not considered part of the dissection.

Number 3

"Of Knights and Cooks, and the Game of Checkers," by Solomon W. Golomb

Abstract

A knight's tour on a standard chessboard is where a knight is placed on a square and then moved so that it visits every square with no repeats. If the knight ends up at the starting square it is known as a closed tour and if it does not it is an open tour.
In this paper a variety of related problems are considered. Knight's tours on rectangles of various sizes and ways in which a set of knights can be placed on a board with no knight attacking any other are examined.
A hybrid game of chess and checkers called "cheskers" that is played only on the black squares is examined. This game was invented by Golomb and a new piece called a "cook" had to be created as a knight cannot travel on this board. The cook moves three squares in one direction and one in the other so that it always remains on a black square. Cook's tours and the placing of non-attacking cooks problems are also considered.

"A Magic Square," by William J. Mannke

Abstract

This paper presents an eight-by-eight magic square constructed by the numbers 1 through 64 where consecutive numbers appear in adjacent squares. In this case adjacent means across, down and diagonally.

"Uncrossed Knight's Tours," by L. D. Yarbrough

Abstract

Knight's tours on boards of various sizes from three-by-three to eight-by-eight where the paths do not cross each other are given.

"Doodling With Numbers," by J. A. H. Hunter

Abstract

This paper contains a set of identities where sums and differences yield the value one and a set of sums where two identical representations in different bases yield a value where the same representation is in a different base.

"Generalized Fibonacci Numbers and the Polygonal Numbers," by V. E. Hoggatt, Jr.

Abstract

The generalized Fibonacci numbers are defined by the recurrence relation

$S(1) = 1, S(2) = 1, S(3) = 2, \ldots, S(r) = 2^{(r-2)}, S(r+1) = 2^{(r-1)}$.

Some power and summation identities for the generalized Fibonacci numbers are given.

The polygonal numbers are defined by the recurrence formula

$P(1,r) = 1, P(n + 1,r) = P(n,r) + n(r - 2) + 1$,

where $r \geq 3$ is the number of sides of the polygon. Additional properties of these numbers are then described.

"Nine-Digit Determinants Equal to 3," by Charles W. Trigg

Abstract

The nine positive digits can be placed into a three-by-three array in 9! ways. It is then possible to compute the determinants of these arrays and the emphasis here is on the arrays where the determinants are equal to three.

"From Forests to Matches," by Ronald C. Read

Abstract

Graph theory is the area of mathematics where figures are made by using a set of nodes and a set of connections between them. The connections can be drawn as either straight or curved segments, depending on the contextual need. The first part of the paper contains some of the basic properties of graphs.

The second part of the paper involves the determination of how many distinct graphs can be created using a specific number of matchsticks, which correspond to the edges. Since the number of nodes can vary, there is a table of (n = nodes, k = edges) values. A large number of figures are included.

"Automorphic Numbers," by Vernon deGuerre and R. A. Fairbairn

Abstract

An automorphic number is a number n such that the end of the square of n is equal to n. For example, 25 * 25 = 625, so 25 is automorphic. An analysis of the structure of automorphic numbers in base 10 as well as other bases and tables of all automorphic numbers up to 1000 digits in bases 6, 10 and 12 are included.

"Division of Integers by Transposition," by Charles W. Trigg

Abstract

There are numbers N such that if the leading digit d is transposed to the trailing end, the result is N / d. Methods for identifying such numbers and a list of some of them in various bases are given.

"Tangrams," by Harry Lindgren

Abstract

The set of seven pieces into which a square can be divided is known as the tans and the set of figures that they can be used to construct are called tangrams. These pieces have been around for some time and in this paper a method of numbering the tangrams for reference purposes is given.

Number 4

"Compounding a Series," by R. Robinson Rowe

Abstract

In this paper, a set of infinite series constructed from the products of the terms of two infinite series are briefly examined. All the series are the infinite sums of the terms of N / n!, where N is a simple series. Two of the examples are $N = n - 1$ and $N = n(n - 1)$.

"Mathematicians and Mathematics on Postage Stamps," by William L. Schaaf

Abstract

Many nations have honored mathematicians and their work via designs on postage stamps and this paper contains a list as well as many quality images of the stamps.

"VF Numbers," by J. A. Lindon

Abstract

VF stands for Visible Factor and a VF number is one where the digits exhibit in a clear way what some of the prime factors are. The most obvious such numbers are those that end in an even digit or 5. The prime factors of more complex VF numbers are recognized by the patterns that they exhibit.

"The Construction of Magic Knight Tours," by T. H. Willcocks

Abstract

A magic knight's tour on a chessboard is one where the moves are numbered 1, 2, 3, ... and placed in the target square as the tour progresses and the result is a magic square. No known magic knight's tour is known, although some semi-magic squares are given. Some of the mathematical techniques used to search for magic knight's tours are presented.

"Recurrent Operations on 1968," by Charles W. Trigg

Abstract

In this paper, several operations are performed on the number 1968 and then repeated several times. For example, the first one is the sum of 1968 and its reversal, where the reversal of that is then added to the sum.

"Reversal Products," by J. A. H. Hunter

Abstract

There are several pairs of two-digit numbers such that the product of the two is the product of their reversals. For example, $(93)(26) = (39)(62)$. The theoretical methods used to identify all such pairs of numbers are given.

"Additional Mathematical Theory of Think-A-Dot," by Sidney Kravitz

Abstract

"Think-a-Dot" is a simple game based on a box with three holes on the top. Each of those holes are connected to a node directly beneath it and the nodes are designated A, B and C left to right. There are two nodes in the row underneath that one, labeled as D and E. The final row has three nodes, labeled F, G and H left to right. The nodes are connected in the following way, A to D, B to D and E and and C to E. D is connected to F and G and E is connected to G and H. Marbles are dropped in the holes and the marble will then follow a path to the bottom row.

There are switches on each connection so that the marbles that reach a node will alternate on which path they take. The states of the switches can be represented using zeroes and ones and the analysis is determining the original state of the switches based on the drops of several marbles and noting the bottom node they end up in.

"Patterns in Primes – Addenda," by Leslie E. Card

Abstract

This paper lists all primes of 2 through 9 digits in length where the digits are consecutive. For the purposes of this analysis 9 and 0 are considered consecutive digits. Snowball primes are sequences of prime numbers where you start with an initial prime and generate additional primes by appending a digit on the low end. A list of snowball primes is also given.

"Strings of 7 and 8 Identical Digits in 2^n," by Edgar Karst

Abstract

A computer program was written to identify power of 2 where there is a sequence of 7 or 8 identical digits. A list of all such numbers for all $n < 100,000$ is given.

Curiosa on 2016 and 2017

Charles Ashbacher

cashbacher@yahoo.com

Abstract

Leon Bankoff was a dentist by profession and a mathematician by choice. Even though he was the dentist for many celebrities, he always found time to do mathematics. His area of expertise was plane geometry and Bankoff served as the problem editor for **Pi Mu Epsilon** from 1968 to 1981. In his paper "Curiosa on 1968[1]" he listed 15 examples where 1968 cubed was the sum of the cubes of three positive integers. This work is a continuation and is dedicated to his memory.

In volume 1, number 2 of **Journal of Recreational Mathematics**[1], a short paper by Leon Bankoff appeared where he listed fifteen examples where 1968^3 was expressed as the sum of the cubes of three positive integers.

In a continuation of that type of work, a search was conducted for instances where 2016^3 and 2017^3 were the sum of three cubes.

The results for 2016 were

$$2016^3 = 14^3 + 994^3 + 1932^3$$
$$= 174^3 + 315^3 + 2013^3$$
$$= 224^3 + 1344^3 + 1792^3$$
$$= 240^3 + 1218^3 + 1854^3$$
$$= 368^3 + 1504^3 + 1680^3$$
$$= 399^3 + 1113^3 + 1890^3$$
$$= 497^3 + 1239^3 + 1834^3$$
$$= 672^3 + 1272^3 + 1800^3$$
$$= 952^3 + 1092^3 + 1820^3$$
$$= 1008^3 + 1344^3 + 1680^3$$
$$= 1296^3 + 1368^3 + 1512^3.$$

The results for 2017 were

$$2017^3 = 876^3 + 1441^3 + 1656^3$$

$= 892^3 + 1185^3 + 1800^3$

Reference

1. L. Bankoff, "Curiosa on 1968," **Journal of Recreational Mathematics**, Vol.1, No. 2, page 78, 1968.

Integer Compositions Applied To The Probability Analysis Of Blackjack And The Infinite Deck Assumption

Jonathan Marino
Roanoke College
jmarino@mail.roanoke.edu

David G. Taylor
Roanoke College
taylor@roanoke.edu

Abstract

Composition theory can be used to analyze and enumerate the number of ways a dealer in Blackjack can reach any given point total. The rules of Blackjack provide several restrictions on the number of compositions of a given number. While theory guarantees a specific number of unrestricted compositions of any positive integer, we must subtract the number of compositions not allowed in Blackjack. We present a constructive approach to enumerate the number of possible compositions for any point value by deleting those illegal compositions from the total number of unrestricted compositions. Our results cover all possible cases and also generalize to changes to the rules of Blackjack, such as the point value where the dealer must stand. Using the infinite deck assumption, we also find the approximate probability that the dealer reaches that point total.

Introduction

In this paper, we explore Blackjack in terms of restricted compositions of the natural numbers. The rules of Blackjack impose several interesting restrictions on compositions of numbers including how to score an Ace as either one or eleven. We provide two ways to enumerate such compositions. One case is a closed form, useful in calculating the number of ways the dealer can reach a number within 11 of his or her face up card, and the other is a general form to calculate all possible situations. Although the closed form is sufficient in most situations, the general form abstracts all of the rules of Blackjack, including the range of card values, the rule where the dealer must stand, and the target total of points.

This pattern was the result of counting the number of ways the dealer could reach a total of 17 points with a face up card of 10 explicitly. If the dealer drew no more cards, there is only one way to reach 17, that is, the dealer's face down card is a 10 (10 + 7). If the dealer drew one more card beyond the face up card and face down cards he or she currently has, there are five ways,

$$10+2+5 \qquad 10+3+4 \qquad 10+4+3 \qquad 10+5+2 \qquad 10+6+1.$$

Continuing to enumerate the possibilities, we have the following pattern

Number of Cards	0	1	2	3	4	5
Ways to Reach 17	1	5	10	10	5	1

which is fifth row of Pascal's Triangle and thus the binomial coefficients for n = 5. We explore why this pattern occurs and generalize our findings.

Partitions and Compositions

Definition 2.1. A **composition** of a positive integer n is a listing $(\lambda_1, \lambda_2, ..., \lambda_m)$ of positive integers such that their sum is n. That is,

$$n = \sum_{i=1}^{m} \lambda_i \text{ where } \lambda_i \in \mathbb{Z}^+$$

where each λ_i is called a part of the composition and the number of parts, m, is the length of the composition. By convention, we may write a composition of n with m parts as $\lambda_1 + \lambda_2 + \cdots + \lambda_m$ or $(\lambda_1, \lambda_2, ..., \lambda_m)$.

Note that, for compositions, the order of the parts is important (the concept of the more-commonly used **integer partition** is simply a composition where order does not matter). The following well-known result will be useful.

Lemma 1. Defining c(m,n) to represent the number of compositions of n with exactly m parts, we have

$$c(m, n) = \binom{n-1}{m-1}.$$

The ability to enumerate compositions as binomial coefficients is key to proving the aforementioned pattern as fact. To aid in counting the restricted compositions, we introduce the following notation.

Definition 2.2. Let $\Lambda(m,n)$ be the set of all compositions of n with m parts. That is, $\Lambda(m,n) = \{(\lambda_1, \lambda_2, ..., \lambda_m) : |\lambda| = n\}$.

Also, it is important to note that

$$|\Lambda(m, n)| = c(m, n) = \binom{n-1}{m-1}.$$

It is also helpful to visualize partitions and compositions as boxes in the form of Young tableaux.

Example 2.1. Let (3,2,4,1) be a composition of of 10 with 4 parts. Then the Young tableau of the composition is

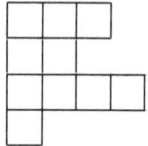

where we note that the top 3 blocks represents $\lambda_1 = 3$, the second row of blocks represents $\lambda_2 = 2$, and so on. There are total of n = 10 blocks, and there are m = 4 rows corresponding to the 4 parts of the composition.

Those familiar with partition theory will recognize Young tableaux. Again, partitions are compositions with the restriction that $\lambda_i \geq \lambda_j$ if $i < j$; that is, for a partition of n with m parts, we have $\lambda_1+\lambda_2+\cdots+\lambda_m = n$ and $\lambda_1 \geq \lambda_2 \geq \cdots \geq \lambda_m$. Young tableaux for partitions have the same typical structure. For example, consider the partition $4 + 3 + 2 + 1 + 1$ of the integer 11. Then we have

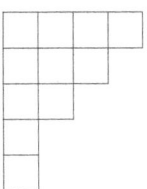

for the Young tableau representing (4, 3, 2, 1, 1). The Young tableau for any given partition will look like an "upside down staircase" due to the restriction that the parts be listed in descending order.

These definitions, along with the following lemma, can be used to help describe our pattern.

Lemma 2 (Variation on Pascal's Rule). For any $n \in \mathbb{Z}^+$, we have

$$\binom{n}{k} - \binom{n-1}{k-1} = \binom{n-1}{k}.$$

Rules and Restrictions of Blackjack

Blackjack has several interesting rules that restrict the number of possibilities of reaching a given point total. Each rule provides a restriction on the total number of compositions that we must remove from the total number, $c(m,n)$. We will start with the total number of compositions of a given number and remove those compositions that are not allowed by the rules of Blackjack.

Since each card has a point value associated with it between 1 and 11, each part of the composition must be less than or equal to 11. Restriction 1 removes compositions with pieces too big to be represented by playing cards.

Restriction 1. For every composition of point values, each part, λ_i must be less than or equal to A, the highest valued card. Note that in standard play, A = 11.

Next, it is usually the case that the first unknown card is not worth one, since if the first unknown is an ace, it would be worth 11. In general, an Ace can only be worth 1 point if the dealer already has more than 10 points, causing the dealer to bust if the Ace is counted as 11.

Restriction 2. The first part of the composition usually cannot be a one, that is, the composition cannot take the form

$$1 + \lambda_2 + \lambda_3 + \cdots + \lambda_m.$$

If the first unknown is an ace, it would be worth 11. In general, an Ace can only be worth 1 point if the dealer already has more than 10 points, causing the dealer to bust if the Ace were counted as 11. In a few rare cases, the first card could be a 1. For example, consider the situation where a dealer with a face up 2 trying to reach 19. Then we allow $(1, 3, 10, 3)$ since

$$2 + (1 + 3 + 10 + 3) = 19.$$

However, this can only happen when the difference between the dealer's goal and his or her face up card is greater than 11. We explore this in detail in the general case of the resulting theorems.

Additionally, the dealer must stand at a certain point value. In standard play, the dealer is required to stop once they reach 17. If the dealer is trying to reach 18, the last card cannot be worth 1, since the dealer will reach 17 in $m - 1$ steps. Similarly, if the dealer is trying to reach 19, the last card cannot be worth 1 or 2, as the dealer would have to stand at 18 or 17, respectively, and be unable to draw additional cards.

Restriction 3. The last part of the composition cannot be any positive integer less than or equal to the difference between the desired number of points, w, and the point value where the dealer must stand, s, that is, the composition cannot take the form

$$\lambda_1 + \lambda_2 + \lambda_3 + \cdots + \lambda_{m-1} + y \text{ where } y \in \mathbb{Z} \text{ such that } 1 \leq y \leq w - s.$$

The previous two restrictions have dealt with Aces at the beginning and end of compositions. Aces can be worth 1 or 11, but the dealer defaults to treating an Ace as an 11. He or she continues to count the Ace as an 11 unless future cards cause the dealer to go over 21; this situation allows the dealer to change the Ace to a 1 and reevaluate the current point total. To account for the behavior of Aces in the middle of our compositions, we introduce the following definition.

Definition 3.1. If d is the point value of the dealer's face up card and s is the point value where the dealer must stand, then we define the i-Ace set to be the subset of all compositions of $s - d$ consisting of all compositions of $A - d - 1$ of length i. That is, the set of compositions of length m of the form

$$\tau_i = \{(\lambda_1, \lambda_2, \ldots, \lambda_i, 1, \lambda_{i+2}, \ldots, \lambda_m) : |\lambda| = s - d, \lambda_j \geq 2 \text{ when } j < i, \text{ and } \lambda_1 + \cdots + \lambda_i \leq A - d - 1\}.$$

Plainly, each i-Ace set represents compositions of $s - d$ of length $m > i + 1$ where part $i + 1$ is a 1 in situations where the 1 must be an 11. The restriction that each part be greater than 2 is to make sure that the $\lambda_{i+1} = 1$ in question is indeed the first 1 in the composition. Symbolically, these are compositions of the form

$$\lambda_1 + \lambda_2 + \cdots + \lambda_i + 1 + \lambda_{i+2} \cdots + \lambda_m = s - d$$

where $d + \lambda_1 + \cdots + \lambda_i \leq 10$. This forces the 1 to be an 11. While these are perfectly allowable compositions of $s - d$, they are not allowed in Blackjack.

Example 3.1. Consider the situation where the player is against the dealer, who has a face up 2 ($d = 2$) and must stand at 17 ($s = 17$). The player is interested in knowing how many ways the dealer can reach 17 with three additional cards (m = 3), including the unknown face down card. We are considering compositions of 15 of length 3. Then the 1-Ace set would be the set

$$\tau_1 = \{(\lambda_1, \lambda_2, \lambda_3) : \lambda_2 = 1, \lambda_{i \neq 2} \geq 2\}.$$

This represents the compositions

2 + 1 + 12	4 + 1 + 10	6 + 1 + 8	8 + 1 + 6
3 + 1 + 11	5 + 1 + 9	7 + 1 + 7	

Note that the first part of each composition is of the form $\lambda_1 \leq 8 = 10-2$. While these are all compositions of 15, we must delete them from our total number of legal Blackjack compositions since, in these situations, it would be impossible that the second card is worth 1 since the dealer would be forced to consider the Ace an 11. However, the composition 3 + 11 + 1 is allowable and not counted in the 1-Ace set.

In our example, we remark that 4 + 1 + 10 is disallowed from our acceptable compositions since 2+(4+11) equals 17 already, so that 11 could not be counted as 1. Also, composition 2+1+12 would be disallowed by our first restriction for ordinary Blackjack where the highest point value for a card is 11. Since the i-Ace sets are placeholders for certain illegal compositions, we must remove them from our total.

Restriction 4. The compositions represented by the i-Ace sets must be deleted.

However, if the composition in question is small enough, we need not concern ourselves with the i-Ace sets. In proving the theorem that explains our pattern by relating the number of allowable compositions to binomial coefficients, the following lemma will be useful.

Lemma 3. Let w be the point total the dealer is trying to reach and d be the dealer's face up card. If the difference between the dealer's goal and face up card is less than 11, that is, $w - d \leq 11$, then $\tau_i = \emptyset$ for all i.

Proof. Let $w - d \leq 11$. Then $w \leq 11 + d$. If $\tau_i \neq \emptyset$, then there exists some composition $\lambda_1 + \lambda_2 + \cdots + \lambda_i \leq 10 - d$. Then we have, $d + \lambda_1 + \cdots + \lambda_i + 11 \leq 21$. But since each $\lambda_i \in \mathbb{Z}^+$, $d + 11 < d + \lambda_1 + \cdots + \lambda_i + 11$, and so $w \leq 11 + d < d + \lambda_1 + \cdots + \lambda_i + 11$, that is, our composition is larger than the target number of points. Therefore that composition cannot exist since the Ace could not be worth 11 points, and so $\tau_i = \emptyset$.

We now have the necessary tools to investigate our pattern further.

Results

If the dealer's face up card has a high enough value, the number of allowable compositions yields a closed form.

Theorem 1. Let s be the the point value where the dealer must stand, d be the point value of the dealer's face-up card, and w be the desired number of points. If $w - d \leq 11$, then the number of ways, n, the dealer can reach a total of w points with m cards (excluding the initial face up card) is

$$n(m, s, d) = \binom{s - d - 2}{m - 1}$$

Proof. We are considering compositions of $w - d$. The total number of compositions of $w - d$ with m parts is given as

$$c(m, w - d) = \binom{w - d - 1}{m - 1}.$$

We will show that by applying the rules of Blackjack, we may remove exactly enough compositions to reach the stated number of compositions. Let R_n be the number of compositions removed by the nth restriction. Then

$$n(m,s,d) = c(m,w - d) - R_1 - R_2 - R_3 - R_4.$$

Note that since $w - d \leq 11$, no composition of $w - d$ can have parts $\lambda_i \geq 12$, and thus $R_1 = 0$. Also, $\tau_i = \emptyset$ for $i = 1, 2, ..., m - 2$ by Lemma 3. So $R_4 = 0$.

Consider Restriction 2. Since the first card cannot be an Ace, we must subtract the number of compositions of $w - d$ that begin with 1, that is, any composition of the form $1 + \lambda_2 + \lambda_3 + \cdots + \lambda_m$. Note that if $1 + \lambda_2 + \lambda_3 + \cdots + \lambda_m = w - d$, then $\lambda_2 + \lambda_3 + \cdots + \lambda_m = w - d - 1$. So these are compositions of $w - d - 1$ with $m - 1$ parts. We can enumerate these compositions as

$$R_2 = c(m - 1, w - d - 1) = \binom{w - d - 2}{m - 2}.$$

Now consider Restriction 3. Since the dealer must stand on s, the last card cannot have a value of between 1 and $w - s$. Thus R_3 must be the number of compositions of $w - d$ of the form $\lambda_1 + \lambda_2 + \cdots + \lambda_{m-1} + y$, where $y \in \mathbb{Z}$ such that $1 \leq y \leq w - s$.

Note that if $\lambda_1 + \lambda_2 + \cdots + \lambda_{m-1} + y = w - d$, then $\lambda_1 + \lambda_2 + \cdots + \lambda_{m-1} = w - d - y$, so R_3 is the number of compositions of $w - d - y$ with $m - 1$ parts. Summing over all of the possibilities of y, we have that

$$R_3 = \sum_{i=1}^{w-s} c(m - 1, w - d - i) = \sum_{i=1}^{w-s} \binom{w - d - i - 1}{m - 2}.$$

However, for each composition ending with y we remove, we have already removed the composition $1+\cdots+y$, so we must add it back. So if $1 + \lambda_2 + \lambda_3 + \cdots + \lambda_{m-1} + y = w - d$, then $\lambda_2 + \lambda_3 + \cdots + \lambda_{m-1} = w - d - y - 1$. Let R^* be the number of doubly removed compositions, that is, the number of compositions of $w - d - i - 1$ with $m - 2$ parts which we must add back. Summing over all of the possibilities of y, we have that

$$R^* = \sum_{i=1}^{w-s} c(m-2, w-d-i-1) = \sum_{i=1}^{w-s} \binom{w-d-i-2}{m-3}.$$

Combining these, we have an adjusted total for the number of compositions of $w - d$ with length m allowed in Blackjack,

$$\begin{aligned}
n(m, s, d) &= c(m, w - d) - R_2 - R_3 + R^* \\
&= \binom{w-d-1}{m-1} - \binom{w-d-2}{m-2} \\
&\quad - \sum_{i=1}^{w-s} \binom{w-d-i-1}{m-2} + \sum_{i=1}^{w-s} \binom{w-d-i-2}{m-3} \\
&= \binom{w-d-1}{m-1} - \binom{w-d-2}{m-2} \\
&\quad - \sum_{i=1}^{w-s} \left(\binom{w-d-i-1}{m-2} - \binom{w-d-i-2}{m-3} \right).
\end{aligned}$$

Using Lemma 2, we may combine the first two terms and simplify the summand, yielding

$$n(m,s,d) = \binom{w-d-2}{m-1} - \sum_{i=1}^{w-s} \binom{w-d-i-2}{m-2}.$$

Expanding the summation, for the right hand side, we have

$$\binom{w-d-2}{m-1} - \binom{w-d-3}{m-2} - \binom{w-d-4}{m-2} - \cdots$$
$$\cdots - \binom{w-d-(w-s-1)-2}{m-2} - \binom{w-d-(w-s)-2}{m-2}.$$

We may again simplify the first two terms using Lemma 2, equaling

$$\binom{w-d-3}{m-1} - \binom{w-d-4}{m-2} - \binom{w-d-5}{m-2} - \cdots$$
$$\cdots - \binom{w-d-(w-s-1)-2}{m-2} - \binom{w-d-(w-s)-2}{m-2}.$$

Again, we may combine the first two terms. With each simplification, the number of parts in the first coefficient is preserved while the next term is still offset by one. This allows us to collapse the sum. We are left with

$$n(m,s,d) = \binom{w-d-(w-s-1)-2}{m-1} - \binom{w-d-(w-s)-2}{m-2}$$

which simplifies to

$$n(m,s,d) = \binom{w-d-(w-s)-2}{m-1} = \binom{s-d-2}{m-1}.$$

Remark 1. The number of possible compositions to reach a total of w does not depend on w at all, but only the dealer's stand rule, s, and the dealer's face up card, d. Since Theorem 1 enumerates the number of compositions as a binomial coefficient, we see that as m increases, we move along the $(s - d - 2)$ row of Pascal's Triangle, agreeing with the aforementioned pattern.

Example 4.1. Consider the situation of a player against a dealer that must stand at $s = 17$ wondering how many ways the dealer can reach $w = 18$ points with a face up Queen ($d = 10$) without drawing any more cards. Then we have

$$n(1,17,10) = \binom{17-10-2}{1-1} = \binom{5}{0} = 1$$

representing the situation that the dealer has an 8 as their face down card.

Example 4.2. Now consider the situation where the dealer has a face up 9 (d = 9). Then the number of ways the dealer could reach a total of 19 points after revealing m = 2 more cards is

$$n(3,17,9) = \binom{17-9-2}{3-1} = \binom{6}{1} = 6.$$

This represents the compositions

$$9 + 2 + 8 \quad 9 + 4 + 6 \quad 9 + 6 + 4$$
$$9 + 3 + 7 \quad 9 + 5 + 5 \quad 9 + 7 + 3.$$

Note that the compositions $9 + 8 + 2$ and $9 + 9 + 1$ would cause the dealer to stand on 17 and 18, respectively, so they are removed from the total number of legal compositions.

4.1. General Case. The hypothesis of Theorem 1 assumed $w - d \leq 11$. We generalize the previous result to allow the dealer to have any face up card.

Theorem 2 (General Case). Let s be the point value where the dealer must stand, d be the point value of the dealer's face-up card, w be the desired number of points, b be the value on which players bust, and A be the value of the highest ranked card. Then the number of ways, n, the dealer can reach a total of w points with m cards (excluding the initial face up card) is

$$g(m,w,s,d) = \binom{s-d-2}{m-1}$$
$$- m \sum_{i=A+1}^{w-d} \binom{w-d-i-1}{m-2} - \sum_{j=2}^{A-d-1} \sum_{t=0}^{m+3} \binom{j-i-2}{i} \binom{s-d-j}{m-3-i}$$
$$+ \sum_{k=b-d-A+1}^{s-d-2} \sum_{i=1}^{m-2} \binom{k-b+s-2}{i-1} \binom{w-d-k-2}{m-i-2}.$$

The extra terms become apparent when $w - d$ is no longer less than 11 since R_1 and R_4 are no longer zero, and a few situations arise where the first card could be a 1. Revisiting our example from earlier, when $w = 19$, $d = 2$, and $s = 17$, we allow $(1,3,10,3)$ since $2 + (1 + 3 + 10 + 3) = 19$.

In some situations, we see that Restriction 2 is removing compositions that should not be removed. This leads us to revisit Restriction 2.

Restriction 2 Revisited. In the general case, an Ace can be the face-down card and count as 1.

It appears that Restriction 2 is removing too many compositions. To compensate for this, we add terms removed from R_2 by noticing that any allowable 1 for an Ace as the face-down card has the form (for this example, $m = 6$)

$$d + 1 + \underbrace{\lambda_2 + \lambda_3 + \lambda_4}_{\text{Piece 1}} \mid + \underbrace{\lambda_5 + \lambda_6}_{\text{Piece 2}} = w$$

where the vertical bar represents the tipping point upon which

$$d + 1 + \lambda_2 < s \text{ and } d + 11 + \lambda_2 + \lambda_3 > b$$

with b representing the largest point total allowed before busting (normally 21).

This algebraically reduces to cases when

$$b - d - 11 < \lambda_2 + \lambda_3 < s - d - 1.$$

We must add back these terms. Let R_2^* be the number of compositions removed by Restriction 2 that should not have been. Then we must add back

$$R_2^* = \sum_{k=b-d-A+1}^{s-d-2} \sum_{i=1}^{m-2} \underbrace{c(i, k-(b-s-1))}_{\text{Piece 1}} \underbrace{c(m-i-1, w-(d+k+1))}_{\text{Piece 2}}$$

$$= \sum_{k=b-d-A+1}^{s-d-2} \sum_{i=1}^{m-2} \binom{k-b+s-2}{i-1}\binom{w-d-k-2}{m-i-2}$$

compositions to our total. Note that this cannot happen when $w - d \leq 11$ so the previous theorem is unaffected.

To aid in proving Theorem 2, we also define a way to enumerate the number of compositions of a given number with each part at least 2. This will be helpful in enumerating the compositions represented by Restriction 4.

Definition 4.1. Let $\hat{c}(m,n)$ be the compositions of n with m parts, with each $\lambda_i \geq 2$.

This definition is helpful in describing the kinds of compositions described by the *i*-Ace sets. In order to be able to count how many compositions Restriction 4 removes, we establish how to enumerate such compositions.

Lemma 4 (Enumerating $\hat{c}$). The number of compositions of n with m parts all at least 2 is the same as the number of compositions of n−m with m parts. That is,

$$\hat{c}(m, n) = c(m, n-m) = \binom{n-m-1}{m-1}.$$

Proof. Let $\widehat{\Lambda}(m, n)$ be the set of all compositions of n with m parts, each at least 2. Define $\phi : \widehat{\Lambda}(m, n) \to \Lambda(m, n - m)$ by $\phi(\lambda_i) = \lambda_i - 1$. Then ϕ takes each λ_i from each composition in $\widehat{\Lambda}$ to $\lambda_i - 1$. Then, since we know that ϕ is a bijection (with inverse map $\phi^{-1}(\lambda_i) = \lambda_i + 1$), $|\widehat{\Lambda}(m,n)| = |\widehat{\Lambda}(m, n-m)|$, and so

$$\hat{c}(m,n) = c(m, n-m) = \binom{n-m-1}{m-1}.$$

The following remark demonstrates the proof of Lemma 4 using Young tableaux.

Remark 2. Consider the composition $3 + 2 + 4 + 2 + 3 = 14$. Note that $(3,2,4,2,3) \in \widehat{\Lambda}(5,14)$. The Young tableau

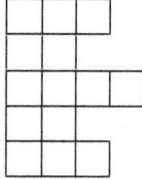

represents the composition $(3,2,4,2,3)$. Then ϕ maps this composition to $(2,1,3,1,2)$ by removing 1 from each piece, as seen in

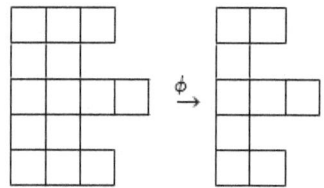

and hence chopping off the first column. Note that the second composition sums to 9, which is $14 - 5$. We describe this action as

$$\phi: \widehat{\Lambda}(5,14) \to \Lambda(5,9).$$

We now have the necessary tools to complete the proof of Theorem 2.

Proof of General Case. We proceed in a constructive way as earlier. For the total number of legal compositions, we have

$$\begin{aligned} g(m,w,s,d) &= c(m, w-d) - R_1 - R_2 - R_3 - R_4 + R^* + R_2^* \\ &= c(m, w-d) - R_2 - R_3 + R^8 - R_1 - R_4 + R_2^* \\ &= \binom{s-d-1}{m-1} - R_1 - R_4 + R_2^* \end{aligned}$$

by Theorem 1. Consider Restriction 1. We must remove all compositions of $w - d$ that have any $\lambda_i > A$. The number of such compositions is equivalent to the number of compositions of $w - d - i$ as i ranges from $A + 1$ to $w - d$. Considering the m possibilities for each one, we have that

$$R_1 = m \sum_{i=A+1}^{w-d} c(m-1, w-d-i) = m \sum_{i=A+1}^{w-d} \binom{w-d-i-1}{m-2}.$$

Now consider Restriction 4. We must remove all of the compositions in each of the i-Ace sets. That is, we must remove

$$R_4 = \sum_{\lambda \in \tau_i} c(m-1-i, s-d-(|\lambda|+1)) = \sum_{\lambda \in \tau_i} \binom{s-d-|\lambda|-2}{m-i-2}$$

compositions from our total. We may represent this as a double summation and expand. Since i ranges from 1 to $m-2$, we have that

$$R_4 = \sum_{i=1}^{m-2} \sum_{\lambda \in \tau_i} \binom{s-d-|\lambda|-2}{m-i-2}.$$

Expanding the outer sum, we have

$$\sum_{\lambda \in \tau_1} \binom{s-d-|\lambda|-2}{m-3} + \sum_{\lambda \in \tau_2} \binom{s-d-|\lambda|-2}{m-4} + \cdots + \sum_{\lambda \in \tau_{m-2}} \binom{s-d-|\lambda|-2}{0}$$

In each sum, we are summing over all of the elements of each τ_i, which ranges from 1 to at most $10 - d$. We manipulate the sums to obtain

$$\sum_{j=2}^{A-d-1} |\{\lambda \in \tau_1 : |\lambda| = j\}| \binom{s-d-j}{m-3} + \cdots + \sum_{j=2}^{A-d-1} |\{\lambda \in \tau_{m-2} : |\lambda| = j\}| \binom{s-d-j}{0}.$$

In essence, we are considering the size of the different i-Ace sets. For each set, we are looking at compositions of j of length $i + 1$ with each part $\lambda_k \geq 2$. This is equivalent to $\hat{c}(i+1, j)$. Using Lemma 4, we change the summands to

$$\sum_{j=2}^{A-d-1}\sum_{i=0}^{m+3} \hat{c}(i+1,j)\binom{s-d-j}{m-3-i}$$

and so, for Restriction 4, we have that

$$R_4 = \sum_{j=2}^{A-d-1}\sum_{i=0}^{m+3} \binom{j-i-2}{i}\binom{s-d-j}{m-3-i}.$$

Combining our results with our revisitation of Restriction 2, we have that

$$g(m,w,s,d) = \binom{s-d-2}{m-1}$$
$$- m\sum_{i=A+1}^{w-d}\binom{w-d-i-1}{m-2} - \sum_{j=2}^{A-d-1}\sum_{i=0}^{m+3}\binom{j-i-2}{i}\binom{s-d-j}{m-3-i}$$
$$+ \sum_{k=b-d-A+1}^{s-d-2}\sum_{i=1}^{m-2}\binom{k-b+s-2}{i-1}\binom{w-d-k-2}{m-i-2}.$$

Example 4.3. We are interested in knowing the number of ways the dealer can reach w = 18 in m = 3 cards with a face up card of d = 2. Plugging in to Theorem 2, we have that

$$g(3,18,17,2) = \binom{17-2-2}{3-1} - 3\sum_{i=12}^{18-2}\binom{18-2-i-1}{3-2}$$
$$- \sum_{j=2}^{10-2}\sum_{i=0}^{3+3}\binom{j-i-2}{i}\binom{17-2-j}{3-3-i}$$
$$+ \sum_{k=21-2-10}^{17-2-2}\sum_{i=1}^{3-2}\binom{k-21+2-2}{i-1}\binom{18-2-k-2}{3-i-2} = 187.$$

Applications in Probability

Now that we can enumerate how many ways the dealer can reach a certain point value, we can calculate the probability that he or she does. For ease of calculations, we introduce the following theorem.

Theorem 3 (Infinite Deck Assumption [2]). The infinite deck assumption in Blackjack fixes the probability of getting any non-ten-valued card as 4 / 52 and getting a ten-valued card as 16 / 52.

From this point on, we will use this assumption. This assumption introduces very little error in calculation, and can be used to determine strategies for games of Blackjack with a large number of decks. Naturally, as the number of decks increases, the error decreases. For a strategy chart for Blackjack with eight decks, only one out of the 290 decision boxes using the infinite deck assumption is wrong, and only by a negligible amount [2].

We combine Theorem 1 with the Infinite Deck Assumption to provide the following proposition.

Proposition 1. Let s be the the point value where the dealer must stand, d be the point value of the dealer's face-up card, and w be the desired number of points. If $w - d \leq 9$, then the probability the dealer's final point total is w is approximately

$$p_w(s, d) = \frac{1}{13}\left(\frac{14}{13}\right)^{s-d-2}$$

Proof. The dealer needs to reach a point value of w in $s - d - 1$ steps or less. Since $w - d \leq 9$, we are only considering cards worth less than 10. If they take one additional card, there are $n(1, s, d)$ possibilities by Theorem 1, each with a probability of $4 / 52 = 1 / 13$, since there are four suits of a card with a particular point value out of a total of 52 cards. Then the probability of reaching w with one additional card, p_w^1, is

$$p_w^1 = n(1, s, d)\left(\frac{1}{13}\right) = \binom{s-d-2}{0}\left(\frac{1}{13}\right).$$

Similarly, if the dealer were to take two cards in addition to the face up card, there are $n(2, s, d)$ possibilities, each with a probability of $\frac{1}{13} \cdot \frac{1}{13}$. If we are considering the composition (λ_1, λ_2), the probability of drawing λ_1 is $1/13$ and the probability of drawing λ_2 is $1/13$. So the probability of reaching w with two additional cards, p_w^2, is

$$p_w^2 = \binom{s-d-2}{1}\left(\frac{1}{13}\right)\left(\frac{1}{13}\right) = \binom{s-d-2}{1}\left(\frac{1}{13}\right)^2.$$

In general, the probability that the dealer's final point total is w after drawing j cards is

$$p_w^j = \binom{s-d-2}{j-1}\left(\frac{1}{13}\right)^j.$$

Since the dealer can reveal anywhere between one and $s - d - 1$ additional cards, we sum over all the probabilities of the allowable compositions to reach w,

$$\sum_{j=1}^{s-d-1} p_w^j = \sum_{j=1}^{s-d-1} \binom{s-d-2}{j-1}\left(\frac{1}{13}\right)^j = \sum_{j=0}^{s-d-2} \binom{s-d-2}{j}\left(\frac{1}{13}\right)^{j+1}$$

$$= \sum_{j=0}^{s-d-2} \binom{s-d-2}{j}\left(\frac{1}{13}\right)^j\left(\frac{1}{13}\right) = \left(\frac{1}{13}\right)\sum_{j=0}^{s-d-2} \binom{s-d-2}{j}\left(\frac{1}{13}\right)^j$$

We may apply the Binomial Theorem to the summation, giving us

$$\sum_{j=1}^{s-d-1} p_w^j = \left(\frac{1}{13}\right)\left(1 + \frac{1}{13}\right)^{s-d-2} = \frac{1}{13}\left(\frac{14}{13}\right)^{s-d-2}.$$

Remark 3. The restriction $w - d \leq 9$ assures us that the probability of drawing a certain card is $1/13$ and that we may use the more attractive closed form of Theorem 1.

Example 5.1. Suppose a player is playing standard Blackjack against a dealer with a face up Jack ($d = 10$). The player is interested in knowing the probability that the dealer reaches 17. Plugging in to Proposition 1 yields

$$p_{17}(17,10) = \frac{1}{13}\left(\frac{14}{13}\right)^{17-10-2} = \frac{1}{13}\left(\frac{14}{13}\right)^5 \approx 0.1114.$$

Example 5.2. Now suppose that the player has 17 points. The dealer, after drawing several cards, has 12 points total. The concerned player is interested in knowing the probability that the dealer beats him. To calculate this, we consider the probability that the dealer ends with a final score between 18 and 21.

$$\sum_{i=18}^{21} p_i(17,12) = \sum_{i=18}^{21} \frac{1}{13}\left(\frac{14}{13}\right)^3 = 4 \cdot \frac{1}{13}\left(\frac{14}{13}\right)^3 \approx 0.3843.$$

Conclusion and Future Goals

We have provided a closed form to calculate the number of ways the dealer can reach a number within 11 of his or her face up card. Although this is sufficient in most situations, we also provide a general form to calculate all possible situations. This general form abstracts all of the rules of Blackjack, including the range of card values, the rule where the dealer must stand, and the target total of points.

In the future, we wish to simplify Theorem 2 to make it into a closed form resembling Theorem 1, allowing us to expand Proposition 1 to include all cases.

References

1. Andrews, G. E. (1998) **The Theory of Partitions**, First Edition, Cambridge University Press.

2. Taylor, D. G. (2015) **The Mathematics of Games: An Introduction to Probability**, CRC Press/Taylor & Francis Group.

BOOK REVIEWS

Edited by:Charles Ashbacher

Charles Ashbacher Technologies

5530 Kacena Ave

Marion, IA 52302

E-mail: cashbacher@yahoo.com

Hidden Figures: The American Dream and the Untold Story of the Black Women Mathematicians Who Helped Win the Space Race, by Margot Lee Shetterly, Harper Collins, New York, NY, 2016. 368 pp., $27.99 (hardbound). ISBN 9780062363596

Given the tremendous military power of the United States in the early twenty-first century and the technical sophistication of the military aircraft, the ways things were decades ago is generally unappreciated. At the start of World War II, like all other aspects of military preparedness, the American aeronautics industry was barely existent. There was a clear need for more and better aircraft, and one of the key skills that was needed was the ability to process numbers.

At this time before the existence of the electronic computer, the term "computer" was generally used to refer to humans that crunched numbers. Most of the time they were female. Yet, this was not a job that required only the ability to punch the right buttons on a calculator, many of the operations required knowledge of advanced mathematics, and people with those skills were in short supply. In the desperate search for talent wherever it could be found, some extremely capable African-American women were recruited. They worked in Hampton, Virginia and at the time that area was subject to strict segregation.

This book follows the lives of four of the women as they started work in World War II and kept working all through the fifties and sixties, playing key roles in the development of new aircraft after the war as well as in the American space program that eventually reached the moon. The four women were Dorothy Vaughan, Mary Jackson, Katherine Johnson and Christine Darden and their experiences with segregation on and off the job make interesting reading.

It is a fascinating book, for it is simultaneously a chronicle of the American war effort and how society changed through greater opportunities as well as a history of the American aeronautic and space program. There is little in the way of mathematics, this is a story of their lives, not the mathematics that they worked on.

To explain how key these women were to the success of the American space effort, all that is needed is one anecdote. In 1962, electronic computers were new and astronaut John Glenn was preparing to be the first American to orbit the Earth. His orbital trajectory had been computed by an IBM 7090 computer, yet he did not trust the results and if they were wrong, he most likely would not survive the flight. Therefore, Glenn uttered what should be one of the most famous phrases of the space program, "Get the girl to check the numbers." He added that if she says the number are good, then he was ready to go.

The person that Glenn was referring to was Katherine Johnson and Glenn, along with everyone else, trusted her. She ran the numbers and they checked out, Glenn was satisfied and his flight was historic. While other events in the space program are not quite so engaging, these women made a difference in the American war and space efforts. They were also wives, mothers and community volunteers, doing all of it while being forced to follow a code that claimed their inferiority. It is a great story of great achievement.

Charles Ashbacher

Algebra for the Practical Man, by J. E. Thompson, D. Van Nostrand Company, Princeton, New Jersey, 1946. 300 pp. (hardbound).

Looking through this book the most obvious feature is that algebra has not changed much in the last seventy years. The only content in this book that is not part of the modern course is the presence of instructions on how to use tables of logarithms.

The book opens with a chapter on the symbols and numbers of algebra, followed by chapters on the rules of addition, subtraction, multiplication and division of numbers. Factorization, powers, roots, exponents, radicals, complex numbers, operations on binomials, solving equations, logarithms, exponential equations, ratios, proportions, progressions, series, combinations and basic probability are the remaining topics. The explanations and examples in the text could be used in modern classes.

Exercises are given at the end of each chapter and solutions to all appear at the end. This is where this book is a refreshing change from the modern textbook. Most of the sets of exercises have only 18-20 problems in them. There seems to be something like an "arms race" among modern textbook authors in terms of how many exercises they include.

Another significant difference is that this book contains no diagrams, it is all text and formulas. It is wise for the educator to take an occasional look back at how it was done many years ago and this book demonstrates the constants inherent in teaching algebra.

Charles Ashbacher

Geeky Lego Crafts: 21 Fun and Quirky Projects, by David Scarfe, No Starch Press, San Francisco, California, 2016. 128 pp. $19.95 (hardbound). ISBN 9781593277673.

The three-dimensional figures that you see in this book range from the nostalgic to those that fit into the holiday mood. My favorite was the set of three alien creatures from the classic "Space

Invaders" video game. All 21 are made from Legos and the patterns are very detailed and easy to follow.

The opening page of the design explanation has a brief section of explanatory text, pictures of the figure(s) and a list of the type and numbers of blocks needed to create each one of them. This is followed by a set of step-by-step visual instructions describing how the figures are created. Given the quality of the explanations for construction, none of the figures is very difficult to build.

The projects in this book would be great for a Boy or Girl Scout project, an activity for school or just something constructive and recreational to do in your spare time.

Charles Ashbacher

The Power of Networks: Six Principles That Connect Our Lives, by Christopher G. Brinton and Mung Chiang, Princeton University Press, Princeton, New Jersey, 2017. 328 pp., $35.00 (hardbound). ISBN 9780691170718.

The two primary uses of the term networking, based on the hardware being used, are covered in this book. When the hardware is human, then the term networking refers to the series of personal and professional relationships that people have. The other use of the networking term is when the hardware is electronic.

As the authors demonstrate, there are many similarities between the two, they often explain a concept of computer networking by using examples of how human relationships function or fail. This brings what often appear to be complex problems down to a level where the non-technical person can understand. The best chapters are those describing the wisdom and stupidity of crowds. Politely stated in sections three and four as "Crowds Are Wise" and "Crowds are Not So Wise."

Many of the fundamental tactics, such as how pages are ranked when searches are conducted, are explained by using a combination of equations and examples between people. The mathematics never rises to a high level of difficulty, if you have had high school algebra you have enough background to understand all but the most complex short sections. It is possible to skip over those sections and still understand the principles being described.

The six broad principles of networking alluded to in the title are:

*) Sharing is hard

*) Ranking is hard

*) Crowds are wise

*) Crowds are not so wise

*) Divide and conquer

*) End to end.

Useful as a resource for learning the basics of networking and much of how social media works, this book could also be used as a supplemental text for courses in the subject. Instructors will find valuable analogies that they can use to explain the concepts in terms that all people can understand.

Charles Ashbacher

Doing the Scholarship of Teaching and Learning (SoTL) In Mathematics, edited by Jacqueline M. Dewar and Curtis D. Bennett, The Mathematical Association of America, Washington, D. C., 2014. 210 pp., $43.00 (print on demand). PDF price $23. ISBN 9780883851937.

For the most part this book introduces a new acronym, but that is the only new concept you will find. Dedicated teachers have been doing what is described in this book since the math teacher position was first defined. Math teachers are always scouring the mathematical countryside looking for new ways to present the subject that they find so fascinating. When something effective is discovered, it is then reported through academic channels.

The titles of the papers in this collection are generally consistent with the titles that have been used in papers about the teaching of mathematics in the past. Some examples of the papers in this book are:

*) "The Question of Transfer: Investigating How Mathematics Contributes to a Liberal Education"

*) "An Investigation Into the Effectiveness of Pre-Class Reading Questions"

*) "Assessing the Effectiveness of Classroom Visual Cues"

*) "Playing Games to Teach Mathematics"

The last item in this list was my favorite in the collection, where the author created a board game using the popular "Trivial Pursuit" game as a model. When this pedagogical technique was used, the inevitable happened, it was better to pit the good math students against each other as well while the weaker ones also played each other. Any veteran of playing the regular game of "Trivial Pursuit" has experienced this, strong players tend to quickly silence the weaker ones.

If you are interested in new ways to educate the latest round of students in mathematics, there are many new ideas in this book. Just don't expect the acronym to be representing anything new under the mathematical sun.

Charles Ashbacher

Magic Square Lexicon: Illustrated, by H. D. Heinz & J. R. Hendricks, published by H. D. Heinz, Surrey, BC, Canada, 2000. 184 pp., $25 (paper). ISBN 0968798500.

This book is the definitive description of the mathematics of magic squares, cubes, hypercubes, stars and other extensions and adaptations. As the title implies, it is in the form of a dictionary and the explanations are complete with detailed illustrations of the items being described. A total of 239 terms are defined, some of which have been in existence for centuries, while others are of very recent origin. Some are original to the authors.

If you are interested in magic squares and the many mathematical and puzzling paths that start at that point, then this is a book that you must read. The authors are clearly experts in their joint labor of love and the result is a significant contribution to an area of mathematics that is accessible to everyone.

Charles Ashbacher

The CS Detective: An Algorithmic Tale of Crime, Conspiracy and Computation, Jeremy Kubica, No Starch Press, San Francisco, California, 2016. 256 pp. $17.95 (paper). ISBN 9781593277499.

This book is different in the sense that it is an unusual combination of what is performed on modern computers with a low technology society with wizards and effective spells. The primary character is Frank Runtime and in a widely used plot device, he is a disgraced former police detective that is now a hardened, cynical private investigator. There is a robbery at police headquarters and Runtime is recruited to hunt the perpetrator(s) down. It is another case where he can do things that are forbidden to the police.

As the investigation continues, there are regular interludes where the algorithmic tactics employed by Runtime are explained to the reader. Most of the algorithms are search algorithms and some examples are breadth-first search, backtracking and binary search. While the explanations are thorough enough to advance the plot of a detective novel, they are not enough to be part of an in-depth education program. There is no mathematics that will be over the head of any reader.

The use of a society where technology is roughly at the middle of the eighteenth century and active magic is practiced is an amusing and unique tactic for detective stories. Therefore, that aspect makes the story entertaining, worthy of reading by people with little to no interest in learning algorithms.

Charles Ashbacher

A Historian Looks Back: The Calculus as Algebra and Selected Writings, by Judith V. Grabiner, The Mathematical Association of America, Washington, D. C., 2010. 287 pp., $62.95 (hardbound). ISBN 9780883855720.

This is literally the reprint of a previously published book followed by some additional articles written by the author. The first section is the book "The Calculus as Algebra: J. L. Lagrange, 1736-1813," that was first published by Garland in 1990. The content is largely as the name implies, a description of how Lagrange worked to treat functions as analytic, so they and their derivatives could be expressed in the form of an infinite series (algebraic expressions). As all students of calculus learn very quickly, a function expressed in this form is very easy to integrate and differentiate. Of course, this was before all was made as rigorous as it is now, so there are some deficiencies in Lagrange's work that are pointed out. Yet, the reader clearly understands the extent of the progress he made.

The remainder of the book contains ten papers by Grabiner, there is no question as to the best, for it answers the question asked of so many calculus students, "Whom do I cuss for this?" That paper is "Who Gave You the Epsilon? Cauchy and the Origins of Rigorous Calculus." It seems a rite of passage for calculus students that they dislike and complain about epsilon-delta proofs. After reading this article, they know who to "blame" for their misery.

In the first paper in the collection of ten, "The Mathematician, the Historian and the History of Mathematics," Grabiner explores the relationship between mathematics, general history and the people that write about their interconnectedness. Most of the major mathematical figures were buffeted by the context within which they lived, so no history of their lives can ever be effectively written without including some understanding of social and political history. This is a thought-provoking paper.

If you are interested in or are teaching a course in the history of mathematics, this is a book you will find valuable.

Charles Ashbacher

The Real Analysis Lifesaver, by Raffi Grinberg, Princeton University Press, Princeton, New Jersey, 2017. 200 pp., $27.95(paper). ISBN 9780691172934.

For most students that take a class in real analysis, there is a rough beginning. They may have struggled a bit with the epsilons and deltas in calculus, but they ultimately triumphed. However, the bar of rigor and detail is dramatically raised in analysis, leading to frustration and uncertainty.

This book is designed to provide the reader some insight into the fundamental material of a real analysis course. As a veteran of that specific war, I certainly recognized the material, the general content was identical to what I worked through. The perspective is different, in the sense that it is a bit more chatty that the usual math book, although the attempts at humor are weak. For example, in the middle of the discussion of the Cantor set on page 119, there is the aside "It is real and perfect (just like you, you special snowflake) . . . "

The explanations are rigid in the sense that no formula is spared, from that perspective it is a typical math book. Detailed proofs of the theorems and lemmas are all included. Yet, there is a lightening of the prose that makes the book more readable than others in the field. It is not a traditional textbook in the sense of containing worked examples or any other exercises. If you are studying real analysis and are looking for another book to read in order to experience a different perspective, this book will serve you well.

From Music to Mathematics: Exploring the Connections, by Gareth E. Roberts, Baltimore: Johns Hopkins University Press, 2016. 301 pp. $47.45 (hardcover), ISBN 978-1-4214-1918-3.

Music and mathematics have been a popular topic for undergraduate students in classes which include an individualized project. As the teacher of such a class, I am familiar with a number of books on this subject. This work by Professor Roberts is a worthy addition to the literature, with a number of distinguishing features. Notably, this is meant to be a textbook for a one- or two-semester course. As such, it includes a nine-page index, exercises and an impressive number of references at the end of each chapter.

The author indicates that his "music first" approach is intended to use students' interest in music as motivation for learning sophisticated and challenging mathematics. I find his exposition of mathematical content to be quite clear and understandable. That being said, this work is also an informative and enjoyable read for the nonacademic reader who may choose to skip some mathematical details.

Chapters 1 through 4 give a thorough treatment of rhythm, scales, chords, tuning and temperament. The physics of sound production, propagation and perception by the human ear-brain system underlie the discussion of loudness and pitch which of course includes a presentation of the sine curve. As an algebraist, I particularly appreciate the depth of the treatment of musical symmetries in Chapter 5, titled Musical Group Theory. True to form, Roberts introduces transformations in music with examples and exercises to familiarize and prepare his reader. Twenty-five pages into the chapter he gives the axiomatic definition of a group, warns the reader that the group operation is not necessarily commutative and then proceeds to the cancellation property, symmetries of the square and finally Lagrange's Theorem. He provides a formal proof of the cancellation property but not of Lagrange's Theorem.

I have found brief mentions of change ringing in other works to be unsatisfying. In chapter 6, Roberts does justice to this topic with details of the history of this art and of the mechanical issues involved in ringing large bells. He shows how physical constraints lead to the "axioms" of change ringing and dictate the permissible patterns. This brings the reader back to more group theory including the symmetric group S_n, the dihedral group D_4 and cosets. The most modern content is in Chapter 8 and includes musical experiments based on magic squares and stochastic mathematics.

I appreciate the author's stated pedagogical objective for this work. My professional opinion as one who has taught both liberal arts courses and upper level math major courses, is that he has provided a book that will allow teachers and students to realize that objective.

Lamarr Widmer

95 Years of Artistic Surprises with
Combinatorial Color-Marked Edge-Matching Tiles
featuring their latest embodiment, MiniMatch-I

by Kate Jones
President, Kadon Enterprises, Inc.

Background

That is a very long title for a little puzzle of just 9 pieces that can form a 3x3 square in 47,563,407,360 different ways, not counting rotations and reflections—

The idea of color-coding the four edges of a square to make them all different yet matchable to their neighbors goes back to a British mathematician, Major Percy MacMahon, and his 1923 book, *More Mathematical Pastimes*. He proposed using three colors taken four at a time to mark the edges of squares, yielding 24 unique square tiles that could form a 4x6 rectangle with all touching sides matching and all outside edges a uniform color.

A few decades later, in the 1970s, an engineer from Lima, Ohio, by name of Wade Philpott, pioneered computer search programs on an antique TRS-80 to identify that MacMahon's 3-Color Squares could have 20 different border configurations and a total of 13,328 solutions. More about Wade's work is here: www.ucalgary.ca/lib-old/SpecColl/philpott.htm and www.gamepuzzles.com/wade.htm

Adding a fourth color raises the number of tiles to 70, fitting into a 7x10 rectangle, again with all matched edges and uniform border color—the Grand Multimatch I *(below left)*.

Fast forward to the Bridges Conference 2015 in Baltimore, MD. The Grand Snowflake, a contoured equivalent of Grand Multimatch I, was chosen for their 2015 art exhibit *(above right)*.

To give visitors an experience of how edgematching works, I designed a miniature set of only 9 pieces (a subset of the Grand Multimatch-I), using 4 colors instead of the dainty contours of the Snowflake set. Leaving the little puzzle open next to the display gave visitors a fun activity with three simple challenges: 1) get all the tiles into the square with matched edges;
2) have one color entirely on the inside, with everything still matching;

3) have two opposite borders a single color, with everything still matching.

At the close of Bridges 2015, further explorations of the little puzzle, now named MiniMatch-I, discovered a surprising richness of variations. Highlights are shown below. All solutions were found by hand.

The designing of MiniMatch-I presented some problems: the four colors filling 36 triangles (4 each on 9 squares) could not be divided evenly into four groups of 9. To match, each color had to occur on an even number of sides. Six of the tiles represented all the different ways the four colors could be placed on the square. Three more tiles were needed, and rather than duplicate one of the six, I added "bowties", where opposite edges shared a color. It took much experimentation to find which colors to use to make the largest variety of goals solvable. So two of the bowtie pieces had 3 colors and the third one had just two. Thus two colors had 10 triangles each, and the other two colors had 8. This allowed those colors to form 5 and 4 internal squares, respectively. More about this set is here: www.gamepuzzles.com/edgmtch5.htm#Mim1

Research Results

- Any one color can be totally enclosed. Several examples are shown below.

- Certain two colors can be simultaneously enclosed (leaving just two colors on the border).

- When opposite borders match at their row ends, the solution "wraps around", forming a cylinder.

- When two opposite pairs of borders match, the wrap-around can go both ways, forming a torus.

- When rows are moved during a wrap-around sequence, the colors and numbers of squares enclosed can change.

- When one pair of opposite borders is mirror-imaged, they represent a Moebius strip.

- When one pair of opposite sides matches and the other pair is mirrored, they represent a Klein bottle.

- When both pairs of opposite borders are mirrored, they represent a projective plane.

- It is curiously difficult to find a solution where none of the 6 rows has matched ends.

- The interior of a matched solution always shows 12 solid-color squares (diamonds). These interior color squares can contain 3 or 4 colors in all these sums:

  5 4 3   5 4 2 1   5 3 3 1   5 3 2 2   4 4 4

  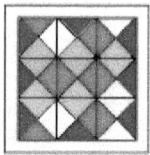   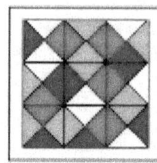 

4 4 3 1	4 4 2 2	4 3 3 2	3 3 3 3
	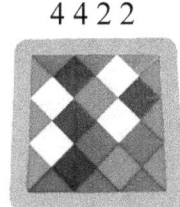	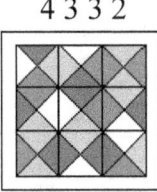	

The remaining two possibilities, 5 5 2 and 5 5 1 1, are not solvable.
- For the special case of 4 4 2 2, any two colors can contribute the 4's, in all 6 possible pairings.
- Only one of the four colors can be entirely on the outside only, no inner square. See 4 4 4 above.
- Besides the 3x3, there are 78 other symmetrical shapes that 9 tiles can form with color-matching.
- Several matched arrangements of the tiles can produce a perfect symmetry of colors.

- Exploring non-match solutions can be very liberating and can also produce color symmetries, an ever changing work of art.

We can only wonder what other amazing surprises this little set will produce. Stay tuned.

MiniMatch and Multimatch are proprietary trademarks of Kadon Enterprises, Inc. See also www.gamepuzzles.com.

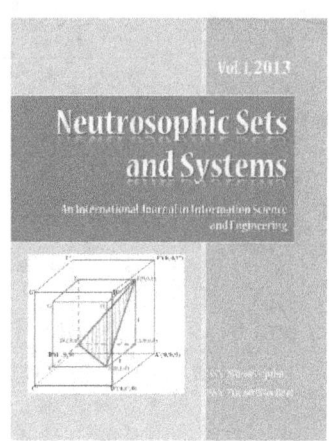

Editor-in-Chief:

Prof. Florentin Smarandache

Department of Mathematics and Science

University of New Mexico

705 Gurley Avenue

Gallup, NM 87301, USA

E-mail: smarand@unm.edu

Home page:
http://fs.gallup.unm.edu/NSS

Associate Editors:

Dmitri Rabounski and Larissa Borissova, independent researchers.

Said Broumi, Univ. of Hassan II Mohammedia, Casablanca, Morocco.

A. A. Salama, Faculty of Science, Port Said University, Egypt.

Yanhui Guo, School of Science, St. Thomas University, Miami, USA.

Francisco Gallego Lupiañez, Universidad Complutense, Madrid, Spain.

Peide Liu, Shandong Universituy of Finance and Economics, China.

Pabitra Kumar Maji, Math Department, K. N. University, WB, India.

S. A. Albolwi, King Abdulaziz Univ., Jeddah, Saudi Arabia.

Mohamed Eisa, Dept. of Computer Science, Port Said Univ., Egypt.

Neutrosophic Sets and Systems has been created for publications on advanced studies in neutrosophy, neutrosophic set, neutrosophic logic, neutrosophic probability, neutrosophic statistics that started in 1995 and their applications in any field, such as the neutrosophic structures developed in algebra, geometry, topology, etc.

The submitted papers should be professional, in good English, containing a brief review of a problem and obtained results. Neutrosophy is a new branch of philosophy that studies the origin, nature, and scope of neutralities, as well as their interactions with different ideational spectra.

This theory considers every notion or idea <A> together with its opposite or negation <antiA> and with their spectrum of neutralities <neutA> in between them (i.e. notions or ideas supporting neither <A> nor <antiA>). The <neutA> and <antiA> ideas together are referred to as <nonA>.

Neutrosophic Set and Logic are generalizations of the fuzzy set and respectively fuzzy logic (especially of intuitionistic fuzzy set and respectively intuitionistic fuzzy logic). In neutrosophic logic a proposition has a degree of truth (T), a degree of indeterminacy (I), and a degree of falsity (F), where T, I, F are standard or non-standard subsets of $]^-0, 1^+[$.

Neutrosophic Probability is a generalization of the classical probability and imprecise probability.

Neutrosophic Statistics is a generalization of the classical statistics.

What distinguishes the neutrosophics from other fields is the <neutA>, which means neither <A> nor <antiA>. <neutA>, which of course depends on <A>, can be indeterminacy, neutrality, tie game, unknown, contradiction, ignorance, imprecision, etc.

All submissions should be designed in MS Word format using our template file:

http://fs.gallup.unm.edu/NSS/NSS-paper-template.doc

A variety of scientific books in many languages can be downloaded freely from the Digital Library of Science:

http://fs.gallup.unm.edu/eBooks-otherformats.htm

To submit a paper, mail the file to the Editor-in-Chief. To order printed issues, contact the Editor-in-Chief. This journal is non-commercial, academic edition. It is printed from private donations.

Information about the neutrosophics you get from the UNM website:

http://fs.gallup.unm.edu/neutrosophy.htm

The home page of the journal is accessed on

http://fs.gallup.unm.edu/NSS

BOOKS IN RECREATIONAL MATHEMATICS BY CHARLES ASHBACHER AND ASSOCIATES

Topics in Recreational Mathematics 1/2015 ISBN 978-1507603215

Topics in Recreational Mathematics 2/2015 ISBN 978-1508617099

Topics in Recreational Mathematics 3/2015 ISBN 978-1511641005

Topics in Recreational Mathematics 4/2015 ISBN 978-1514317518

Topics in Recreational Mathastics 5/2015 ISBN 978-1519115676

Topics in Recreational Mathematics 1/2016 ISBN 978-1530003655

Topics in Recreational Mathematics 2/2016 ISBN 978-1534964846

Topics in Recreational Mathematics 3/2016 ISBN 978-1537333212

Alphametics as Expressed in Recreational Mathematics Magazine ISBN 978-1508538134

Ten Year Cumulative Index to the Journal of Recreational Mathematics, edited by Joseph S. Madachy and Charles Ashbacher ISBN 978-1508936800

Alphametics Expressing Thoughts From the Star Trek Original Series ISBN 978-1512152784

Mathematical Cartoons ISBN 978-1514207130

Solved Problems in Statistical Inference ISBN 978-1515215622

Table of Contents Pages From all 38 Volumes of Journal of Recreational Mathematics ISBN 978-1539652588

Associates

Artist Caytie Ribble

Editor Rachel Pollari

Editor Jennifer Corrigan

Artist Jenna Richardson

www.ingramcontent.com/pod-product-compliance
Lightning Source LLC
Chambersburg PA
CBHW062357220526
45472CB00008B/1834